NORTH STAR

A NAVIGATOR'S ANGLE

DAVID WELCH

SunRise

SunRise

First published in Great Britain in 2023 by SunRise

SunRise
A Division of Upper Octave Ltd
124 City Road
London EC1V 2NX

ISBN 978-1-9144894-0-2

A CIP catalogue record for this book is available from the British
Library.

Typeset in Minion Pro and Impact.

The author has donated his royalties to
The Red Cross Ukraine Appeal

https://donate.redcross.org.uk/appeal/ukraine-crisis-appeal

Contents

Foreword

The 1980s was, for many, a memorable decade, with outlandish fashions, the arrival of the mobile phone, the Filofax and of 'yuppies'. For some, like myself, those years heralded new opportunities in the world of aviation and marked the beginnings of full, exciting and hugely rewarding flying careers.

Much to my headmaster's chagrin, many of my formative days were spent at Wellesbourne Mountford Aerodrome a few miles outside Stratford-upon-Avon. Once the home of a Wellington Operational Training Unit and later the RAF's School of Photography, the site had been closed and sold back to its pre-War owners in the 1960s only to re-opened in 1981 as a licenced civilian airfield.

Working in the tower, cleaning private aircraft in return for the occasional flight, supporting the Airfield Manager as his 'odd job man' and eventually gaining my flying wings, I became addicted to being amongst a wonderfully diverse resident group of aviators, both amateur and professional. They were truly influential and very happy days.

In the respected vanguard (no pun intended) of the professional flyers at Wellesbourne was the author of this book, Neville David Welch (NDW). A no-nonsense, vastly experienced, and often irreverent individual who possessed a gift for imparting knowledge and encouraging others to achieve their goals. When 'Welch' arrived in the Flying Club or the airfield's 'greasy spoon', people sat up and noticed! He was a 'real' pilot, and instructors, the self-improvers in training as well as those who worked in ground trades gravitated towards him in the hope of hearing some of his wise counsel.

Drawing on an extensive and unique background in the British and New Zealand Armed Forces, general aviation, the international airlines and regulators, such as the UK's Civil Aviation Authority, NDW had—and still has—much to offer those willing to learn and improve.

In the cockpit, he had a pretty laid-back and low-pressure instructional technique, but he was consistently quick to spot errors and correct them in a courteous and easy to understand manner. Whilst he didn't instruct me as I worked towards my Private Pilots' License, he and I subsequently flew together several times from Wellesbourne in both the Club's fleet of Cessnas and Pipers but also in more exotic types like Tiger Moths, his own Luscombe and J3 Cubs. Later, he gave this young fighter jock an experience as loadmaster-in-training aboard a Handley Page Herald on a return night trip from Coventry to Ostend; he tells me that he wanted to confirm that I had chosen the right path in joining the RAF!

As you will read in the pages of this book, NDW has had an adventurous life, operating around the world with a host of 'characters'—many of whom had early careers in the Royal Air Force's Bomber Command—and in some classic aircraft types, often in situations that would now be seen as high risk by more faint-hearted and auto-pilot reliant beings.

He has ranged around the seats of the cockpit, at one time being one of the select few airline navigators cleared to operate 'over the pole' flights in the Arctic. And he has the distinction of being one of very few civil aviation pilots licensed both as navigator and ground engineer.

Despite giving his life to the aerospace sector, NDW has always found time to pursue many other interests including New Orleans Jazz, Bach church organ music and the Goon Shows of the 1950s (for aficionados, he still calls me 'Jim'). He is a stalwart of *Air Britain*, has and encyclopaedic knowledge of de Havilland Moths and he still travels the world every year to record rare and early aircraft.

He is a legend to those who know him, and I, together with countless others, owe him a great debt of gratitude. I am delighted that he has chosen to share his reflections with us.

The words of John Gillespie Magee's poem *High Flight* are often used to describe the feelings of an aviator. His haunting lines are incredibly personal to each of us who has had the privilege of enjoying lives in the air. I sense that NDW, more than most, has 'joined the tumbling mirth of sun-split clouds—and done a hundred things you have not dreamed of'.

Air Marshal (retired) Sir Gerry Mayhew KCB CBE

Chapter 1 - Early Years

I think I was conceived on the day war was declared. When I was born, my father was missing in action and my mother lay in hospital with puerperal fever. The wards were overflowing with evacuees from Dunkirk, and her doctors thought she would die. One of her cousins offered to adopt me but my mother defied medical opinion and recovered. My father also beat the odds when he returned from occupied France and arrived in Plymouth; starving and exhausted.

As a military policeman, he had been given the job of directing traffic and was one of the last to leave. When recovered, he was posted to Singapore but while on embarkation leave that October—although I cannot remember it—we saw a German bomber appear over Stafford in broad daylight. It was attacking the English Electric works but did a poor job. One bomb hit a tank production shop but failed to detonate and the other two merely punched holes in the sports field. Meanwhile, we were down in an air raid shelter trying to 'Keep Calm and Carry On'. Many years later I learned that a Luftwaffe bomber was shot down near Crewe, Cheshire, on the same day. The Junkers 88, captained by Oberfeldwebel Heinrich Matzel, was probably our attacker.[1]

The town almost had another Luftwaffe visit in 1944 when an air launched V-1 doodlebug—aimed at Liverpool—blew up in a farm field about ten miles to the west.

---

1 Ju88A-1, serial number 8039, coded B3 + KH.

All I remember was the staff in the day nursery talking about it the next day. As a small boy I didn't think too much about the war, but I knew the bombing was meant to kill us, and it gave me nightmares. Apart from that, I can only remember flashes of light in the night sky which must have been anti-aircraft flak over Birmingham and the Black Country to the south.

Lots of Italian prisoners of war were 'trusties' working in the nearby timber yard. We could tell who they were from the large circular coloured identification patches sewn on the back of their overalls. I guess they were glad to be out of the war—they all seemed to want to make a fuss of local children—but whenever one tried to lean over the garden fence to make a fuss of me, I'd recoil in horror and run. I'd been told they were bad men!

Mother, Aunt Betsy and I took a week's holiday in Rhyl, North Wales in 1943. Although the train trip—changing at Crewe—was not all that long, it seemed to take ages to me. Trains were very crowded during wartime so much of the trip was spent sitting on our suitcases in the corridor, and at night all the blinds had to be down because of the blackout. That was where I saw my first real Spitfire, parked on the promenade as part of a 'Wings for Victory' appeal. Other memories of that wartime holiday were getting badly sunburned and then stung by a jellyfish.

The barrage balloons over Crewe railway junction and marshalling yards were an impressive sight. The route went past a couple of active aerodromes, but at that age I was more interested in trains. Towards the end of the war the nearby Seighford aerodrome was used to train glider pilots to replace the losses at Arnhem. Seeing a huge bomber towing an equally massive glider was impressive and the whole street usually turned out to watch.

Although he had retrained as a military policeman, Dad originally joined the Army as a medical orderly and served for four years in Palestine before the war, seeing considerable violence between Jews and Arabs. Hoping to go back there, he'd swapped postings with another soldier of the same rank and trade while passing South Africa in a

Earliest form of transportation in late 1940. That's where I remember seeing the flak bursts over the Black Country.

troopship. My father went back to the Middle East, but the other man almost certainly became a victim of the Japanese.

Like so many veterans, he never spoke much about his wartime experiences although he did mention he had witnessed the sinking of the *Lancastria*, in which about a thousand RAF ground crew were killed. Military Police had the unsavoury duty of having to arrange and supervise firing squads. He told me he had been involved in the execution of some Basuto soldiers who had broken out of camp, murdered an army sergeant and gangraped his girlfriend, an army nurse. I provoked him into opening up about that, because I had first heard the story from a friend who had been there in the Royal Army Medical Corps. I'll never forget Dad's shocked look when I just happened to mention I knew something about it. Another time, he travelled to Kufra Oasis in the far south of Libya with the Long Range Desert Group. He never told me why, and only mentioned it when I happened to say I'd often flown over Kufra.

He was lucky to have survived the war. He'd been on duty, riding a motor bike, in the Canal Zone when a truck ran him down in the blackout. Dragged for a couple of miles under the truck, he survived with multiple fractures of pelvis and skull and numerous other injuries. Hospitalised for nearly a year in Cairo, he somehow pulled through but died of a brain tumour at the age of fifty-nine. The Dunkirk Veterans Association tried to get my mother an army pension, claiming that the tumour was caused by the accident, which had damaged his skull, but the army produced all his medical records back to the day he joined and convinced the court that the conditions were not associated.

He only flew four times in his whole life. The first was travelling in a Royal Air Force Lockheed to attend a court martial as a witness. He had dysentery at the time and spent most of the journey on the 'john'. The next time was nearly forty years later with the airline I worked for. I arranged what was called a 'Special Passenger Advice', and he was invited to visit the flight deck when the rest of the family took a holiday in Majorca. I couldn't stop him talking about that for months.

He did say he'd spent a lot of the time with a Special Investigations Branch (SIB) unit doing undercover work tracing spies, deserters, drug dealers and, most of all, searching for the Grand Mufti of Jerusalem—the Arab leader who founded Hamas. However, the British never did find him; it turned out he was living in Berlin as a guest of Adolph Hitler. Dad also spent some months as a member of General Montgomery's personal security squad. He opened up about that when I happened to mention I'd been staying at the Mena House Hotel near Cairo; the location of the Cairo Conference attended by 'Monty' and Churchill, in 1943.

At one point he had been invited to transfer to the RAF as an air gunner, since some were being trained locally in Egypt. That would have meant a first tour with one of the bomber squadrons based in the Middle East, and the possibility of a home posting for a second tour. Since air gunners had a low survival rate, especially in Bomber Command, it's probably just as well he didn't accept.

My mother was the eldest daughter of an express train driver with a very large family and although she always

On spotting this on the mantel shelf, a lady visiting my mother remarked, 'Oh! What a beautiful little ... Oh my God, NO !'

kept in touch with her natural parents, she had been adopted by an elderly couple with whom the family had once lived. Curiously, they were brother and sister, but neither had ever been married. Uncle George was a retired steam roller and traction engine driver and Aunt Betsy had spent her career in domestic service as a housekeeper. Both were deeply religious. We also had a lodger in our tiny two-bedroom cottage, an elderly gentleman who worked as a pattern maker at Bagnall's Locomotive Works, nearby.

As was quite usual in working class circles during the war, the two old gentlemen had to share a double bed in the back room. Aunt Betsy and my mother shared a bed in the front room, and I slept in a cot in the corner. So sad that after lifetimes of hard work, the only pleasures those old dears seem to have had was a jug of beer with some cheese and pickled onions on a Saturday evening. Most of their spare time was spent reading the newspapers or, in Uncle George's case, taking me for long walks.

Our lodger, Tommy Dobbin, was taken ill in 1942 and admitted to hospital. My earliest recollection is being taken to visit him. He gave me a police whistle, with which I made so much noise around the cottage that it was very soon confiscated. Sadly, Tommy passed away, and I clearly recall being taken in my pushchair to the funeral in Stafford Cemetery. Nearby, a bugler was sounding the *Last Post* during a military funeral. The parson quoted Isaiah 22:13, 'Let us eat drink and be merry; for tomorrow we shall die.' Ironic advice in the middle of wartime food shortages.

Sadly, dear Uncle George also passed away in late 1943 and I am sorry that I have no photograph, just vague and blurred memories of a kind, gentle, snow-white-haired old gentleman. My mother told me that in his youth he had witnessed one of the last public hangings outside Stafford Gaol.

My boyhood interests had been much the same as any other youngster: railways and models, woodwork, bird watching and astronomy. Apart from scraping through the Eleven-plus, the first exam I ever passed was for the issue of the Starman Badge in the Sea Scouts, but eventually I was chucked out for lack of interest in sailing boats, rigging types, and bosun's whistles. Sixteen years later I would be making my living peering at the stars though a periscopic sextant.

My known forebears had mostly been policemen, army men, or worked for the LMS Railway. Although mother had dated an RAF Pilot before she met my Dad[2], I'd had no contact with anyone who knew much about flying and only knew that 'pilots can't wear glasses'. Becoming a mechanic would have been enough. I'd never suggested I wanted to be a pilot, and kids on the council estate we moved to later didn't then have that level of ambition.

With my father in Egypt and Palestine we were effectively a single parent family for the next five years, so I grew up, like so many wartime kids, not entirely understanding what a 'Daddy' was. To this day I can remember the summer evening when he arrived home after five years in the Middle East. Everything was fine for the first few months but after that we never really got along very well until I was almost thirty. He had grown up after World War One without a father, so hadn't much idea how to handle a small boy, especially a smart alec with no interest in football or cricket.

For some months after the war, Dad remained in the

---

2   Flying Officer Fred Ridley of 105 Squadron was killed in 1940 flying a Fairey Battle in a bombing attack on a bridge near Sedan.

military police, stationed near Chester, with an undercover Special Intelligence Branch (SIB) unit. When he came home, he wore civilian clothes and drove a huge car. What impressed me more was the huge revolver he always carried. He was offered a commission but that would have meant accepting a posting to Egypt again, and leaving Aunt Betsy alone was out of the question. She was disabled by a stroke and had no surviv-

Boy racer terrorising the pedestrians in Stafford Victoria Park, 1943.

ing relatives, so he resigned and joined the civil police as a constable. After a few years, he left and took over as chief of the local Ambulance Service. We moved to a nice modern council house that went with the job. The downside was that the steel framed, single glazed windows made the house freezing cold in winter. That was no problem for anyone else in the street for they were all coal miners and got a free issue of coal, but they were kind people who would slip us the odd bucket from their unwanted surplus—illegal, but much appreciated.

Both my parents were keen dancers and talented swimmers, Dad sometimes played water polo. One day I fell in the deep end and nearly drowned before he could get to me. That put me off swimming for years and although I tried hard enough in my twenties, I could never get it quite right. Maybe I have a mild hand–foot coordination problem because I was never able to learn to dance either.

Maybe just too big and clumsy. Another thing my dad was pretty good at was conjuring tricks. He could pull all those corny old stunts like suddenly yanking a boiled egg out of his ear. I always wished he would show me how, but he never did. That sort of talent can be useful socially!

The old cottage where I was born. That's Aunt Betsy in about 1900.

Chapter 2 - Aeroplanes

I came to flying through porridge. In 1952, my mother found a picture card of an Avro Vulcan in a packet of Mornflake Mighty Oats. Unaware that flying is an incurable addiction, I began to collect the set.

Aunt Betsy had been ill for some time and had died in the previous year. We took in a lodger, a young doctor who was joining a local practice as junior partner. Peter was quite a character and he stayed with us for about two years. He had some strange habits though. For one thing, he didn't think it was proper for the local doctor to be seen in the off-licence buying bottles of beer. We had a deal whereby I would take his brown ale bottles back and collect the new consignment while he sat in his car over the road. My cut was all the pennies for the returned empties, but I don't think the subterfuge fooled anyone for his black Ford Zephyr was the biggest car in the district and always parked on our front driveway. Peter was what would be known as a 'Petrol Head' these days, and often went to car race meetings. He took us all the way to Thruxton Raceway one day, but I was rather disappointed there were no aeroplanes to be seen on the old airfield. Then, out of the clouds there suddenly appeared a massive six-engined USAF Boeing B47 Stratojet bomber, probably on approach to Boscombe Down aerodrome. It was the first one I'd seen but none of my school chums would believe me!

I got into big trouble one evening when I was about ten. I'd been out in the fields playing with some friends, one of whom was very liberal with a certain four-letter word I had not heard before. As dusk fell, I wandered home to find my mother entertaining a couple of her very posh friends with coffee and cakes. I had a little school homework to catch up with and I was searching all over for a ruler. Mother asked what I was looking for and had the reply 'I can't find my fucking ruler mum!'

I will leave it to you, gentle reader, to imagine the reaction, but Krakatoa and Mount St Helens were nothing in comparison. For many days I would have been more comfortable eating my meals off the mantel shelf.

The picture cards became a full set, which I still dig out and look at now and then. Then I discovered that the small aeroplanes that often flew over our house in Staffordshire, came from Wolverhampton Airport, about ten miles away. One day in May 1954, I borrowed Dad's bicycle and set off to find them.

About a mile short of it, I asked another lad the way and it turned out that he was on the way there himself, for his father owned a Miles Messenger based there. He was very kind indeed and showed me all around the five hangars which were chock full of various De Havilland, Miles, and Auster types. A little older than me, he had clearly done a fair amount of flying himself and was a mine of information explaining a lot of things. Pity I never kept in touch with him. All I remember is that his name was Alan—I think. I especially recall a line of Tiger Moths in RAF colours, which were awaiting their turn for some work in the hangar, before being ferried to Thruxton where most of them were rebuilt as four seat Jackaroos.

On of my friends at the time was 'Specky' Davies, son of the local shoe repairer. Why he had that nickname I know not. He didn't wear glasses, but I did. Specky had collected a huge box of newspaper cuttings about aircraft and seemed to know a lot about them. He came with me on a few bicycle expeditions in search of aeroplanes, but his interests began to diverge when he took up the violin. Sixty years later I learned that Professor Howard Davies had become one of the top international concert violinists, leading the world-famous Alberni String Quartet for thirty-five years. Sadly, he had passed away by the time I discovered this.

My father as a Staff Sergeant, Royal Military Police, Egypt 1943.
I'm sure I never looked that smart in uniform.

As time went by, I became more adventurous and ranged far and wide on Dad's bike, taking in Derby (Burnaston) airport, just as all the Spartan Air Services Mosquito bombers were being converted to make the first full photographic survey of the Canadian North. I also went to Birmingham (Elmdon) which in those days only had about five airline movements a day. Unfortunately, that was about the time that flying ceased at the RAF navigation school at Lichfield (Fradley), Walsall civil airport, and at RAF Bobbington. At Derby, I met up with an older chap who arrived on an ancient motorbike and suggested that I take photos instead of just logging registrations. He was Charles Waterfall, who, along with Neville Franklin, was co-founder of the Newark Air Museum. Our friendship lasted until Chas passed away in 2004, after which I bought his negative collection from his widow, Madge. I have to thank Chas for reinforcing my interest, which in time led to a rewarding career in aviation as well as a current collection of about 16,000 antique aircraft images. Had it not been for him, my interest might have lapsed, and my life taken a completely different course.

My mother had a 1930 vintage Kodak Box Brownie, which took Kodak Verichrome Pan film, at 620 size. With eight pictures on a roll, it produced contact prints 4cm x 6cm. I still have some of the negatives and prints but my pocket money restricted how much film and processing I could afford, so many subjects had to be passed over.

On weekends, RAF stations were not very busy so my old school mate Geoff Steele and I would sometimes play truant from school, touring around Shawbury, High Ercall, Cosford and Ternhill. Of course, we saw many hundreds of very interesting types but—without binoculars—were not able to log many serial numbers. With only rudimentary cameras we rarely got close enough to take any good pictures and quite often we were ordered to clear off by the MoD Police guards. At RAF High Ercall, I especially remember a hangar crammed with late mark Spitfires. I don't think many of these were scrapped though. Some were later issued as 'Gate Guards' and still survive to the present. One was TD248 which, of course, is now flying again. There were fields filled with hundreds of surplus Percival Prentice trainers which had been sold to airline entrepreneur Freddy Laker. In the following weeks, we would sometimes see loose formations of about five of them being ferried to Southend and Stansted. Unfortunately, that doesn't seem to have been one of Mr Laker's better schemes for very few were ever converted for civil use and most went to the smelter.

Battle of Britain Air shows were a 'must' in those days. My first was a murky day at RAF Castle Bromwich in 1954. I logged most of the numbers, but that was before I began photography. 1955 saw us at RAF Fradley, which was a very good show in

My father as RSM of a Royal Military Police unit in about 1955.

fine weather. The trouble there was that word had got out that photography was banned—although that may have just been a rumour—so as relatively law-abiding citizens we didn't take our cameras. At Fradley I closely missed having my first flight, which would have been in the last Percival Q Six airliner, G-AEYE, a beautiful blue and white aeroplane flown by Geoffrey Allington, the famous test pilot. We had just reached the top of the waiting line when the air show began and pleasure flying stopped. By the time it started again, we were so far down the queue that it was time to leave for the twenty-mile cycle ride home. In 1956 I was at Cosford and in 1957 back at Castle Bromwich. I still have a few photos, but most were contaminated by people getting in the way and I am still looking for an effective 'human filter'.

Chapter 3 - Schooldays

At Grammar School, I was actively discouraged from my interest in aviation, since it was seen to be affecting my schoolwork. Wearing glasses, I thought I had no chance of getting to fly for a living and a career in aircraft engineering was impossible unless I did much better in maths and physics and got a university place. This was nonsense of course, all I wanted to be was a 'hands on' licensed mechanic and I suppose it shows how unworldly our careers master was. I used to do quite well in history, geography and English and was encouraged to aim for a career as a lawyer, despite having no interest in English literature, Latin or French. Additionally, my parents could not have afforded to get me a junior place in a law office at that time. Despite that, I have always had a great interest in civil law and even put in a lot of effort to get an external law degree in my early thirties. That ambition, however, would be placed on the back burner when I decided to retrain as a pilot, and I regret that it is still there. The biggest distraction from schoolwork, however, wasn't aviation but a delectable little blonde about three years my junior, but that also came to nothing.

I'm afraid languages were always my weak point but, in the countries I visited as aircrew, the only one I thought could be any use was Arabic. Dad spoke it pretty well fluently, although he couldn't read or write the 'barbed wire'. I had often asked him to teach me some, but he always said, 'Never mind flippin' Arabic, get on with your French homework'. Much later I made some effort to learn Dutch but the problem with that was the Dutch themselves. Nobody wanted to let me practice on them. If I said anything in Dutch, they'd always answer me in English, but I suppose that shows they could just about understand me. One captain even described it as a dead language, so they all wanted to use me to practice their English.

The brains behind the family. My mother in about 1960.

There had been one huge bonus during my first year at Grammar School. That came in the form of a temporary science teacher nicknamed 'Walrus' thanks to his massive moustache. Mr Chalmers spent the summer term science lessons teaching us all about the workings of the internal combustion engine. I'm sure it wasn't included in the GCE physics syllabus, but it was more use to me than a lot of the other stuff. We kids liked to imitate his pronunciation of internal com*bustion* engine.

The headmaster was a very popular and respected Lieutenant Colonel Herbert Heatley Hutchinson MC, a kindly First World War veteran and classics scholar. Unfortunately, he took early retirement at the end of my first year in protest against the admission of girls. His replacement was a very gaunt looking character who—for some obscure reason—was nicknamed the Bong. One lunchtime there was suddenly some commotion around the geography room and within minutes all of us

With my Dad on Beachy Head in 1948. I think that's an old RAF radar in the distance.

were herded into the assembly hall. Hushed silence was complete for once, for already the rumour had spread. The Bong appeared in his chalk dusted black gown and, after an interval of a few seconds, announced in funereal tones that a boy had died. We were all sent home for the rest of the day.

The lad had been horsing around with his friends when suddenly his face lost colour and he collapsed and died. I was rather sorry for the Bong. Only a few days before the two of us had been waiting outside his study to receive 'six of the best'. I can still remember that boy asking me if I thought he'd put enough pairs of football shorts under his trousers. Must have been first time for him—I never bothered with extra layers—as one of the Bong's regular clients I was quite inured to it. If central casting had required someone to play the Grim Reaper, we surely could have offered the Bong!

I'm amazed that bullying is still apparently endemic in schools. There was a lot of it in the fifties and kids with glasses tended to get picked upon. For some time, I put up with harassment from a bigger kid about two years older than me. Then one

Smart Alec of the Lower Fourth. Rugeley Grammar School, 1955

day he came up behind me and shoved a handful of dirty puddle water in my face. I went for the blighter and nearly slaughtered him. Of course, when the dust cleared, we both got the whack for fighting, but he kept his distance after that. Teachers were not much help either. All you ever got if you complained was 'You should stick up for yourself boy'. Yes, and see what happens if you do!

There was some light relief though. Two years ahead of us was one Aubrey Watmough (not his real name, of course). Watmough was very studious and an accomplished sportsman. Consequently, he was held up as an example to the rest of our year, until we were sick of hearing about all his wonderful accomplishments. He came over as rather surly and a bit of a creep anyway.

So, one day the teacher failed to appear, and our classroom was in the usual uproar despite the efforts of patrolling prefects. It appeared that an emergency staff meeting had been called. Then a kid from a lower class poked his curly head and freckled face round the classroom door and breathlessly shouted in his dulcet 'Black Country' accented tones, 'Ave yo 'eard about Watmough? His girl-friend's 'aving a babby! He's fooked 'er, fooked 'er 'e 'as!' He then rushed off to spread further the glad tidings. Seconds of stunned silence were followed by hoorays and howls of hysterical laughter. In due course, normal lessons resumed and nothing further was said, but Watmough didn't make his routine appearance at the annual prize giving and was never seen again. That was a great pity for he had assumed an honoured place in the pantheon of our heroes. If we could have named a constellation after Watmough we surely would have done so.

That was about the time I discovered, or perhaps invented, the 'Wurtle'. This was a cartoon like animal which evolved in a strange manner. Doc Peter had been explaining the workings of the female reproductive parts to my mother and had drawn a sketch too late to realise what he'd sketched upon could not be thrown out. So, Peter altered it to look rather like a rabbit. I got hold of it— unaware of its origin— and altered it again. A few of these cartoon animals which, for some reason, I have forgotten, became known as Wurtles and began to appear in some of my school exercise books. Always good for a few laughs, my colleagues soon battened on to the idea and the school became infested with a plague of Wurtles. Eventually, the headmaster made an announcement in morning assembly that this had got to stop. There were some furtive glances at me—blushing bright—but I knew that some of the staff were in on the joke and sniggering behind his back.

Eventually, I was grudgingly permitted to join the Air Training Corps (ATC). My father—now the Regimental Seargent Major of a Territorial Army Military Police unit—thought it would instil some discipline and get me ready for National Service in the army. The ATC is a wonderful organisation and, in my view, worth every penny the Ministry of Defence spends on it. Not just as a service pre-entry organisation either. As with the other cadet forces, it strives to train kids to be good citizens and stay out of trouble. One or two nights a week taught me more about life than six years frittered away on Dickens, Shakespeare, and Latin declensions at Grammar School. In the fifties, grammar schools seemed to be largely staffed by pompous, talentless people with non-vocational degrees, and little aptitude for teaching anyway. Even the school 'Careers Master' seemed to know about little other than teaching French or joining the forces. Mind you, it can't have been easy for some of them. Like 'Boris', our woodwork master who, thanks to timber shortages, was trying to teach us woodwork without any wood. The George Bernard Shaw quotation, 'Those who can, do, do; those who can't, teach.' surely applied to most of our teachers.

Our Latin and physical education teacher was an ex-RAF flight engineer who taught us Latin with a Cardiff accent. At least 'Mel' had a sense of humour. One example chalked on the blackboard translated, 'The wounded Caesar was carried into the fish and chip shop'. I used to do a virtuoso imitation of his voice, when he would bound into the changing room loudly enquiring, 'Right, now where are the non-combatants, publicans and sinners?' This usually meant my chum Johnny McHale, who'd managed to wheedle an 'excused chit' out of his mother; an acute case of 'sniffles' or whatever. Many years later, I learned that Mel had once parachuted out of a Lancaster bomber over Bremen and spent the remainder of the war in a prison camp.

One teacher did stick out head and shoulders above the others. A Miss Stanway took my class for mathematics for my final two years. She was a brilliant, no-nonsense teacher who explained every new topic meticulously before giving us any practice examples. The chap we had before her seemed to spend half his time analysing the previous sports afternoon's football match. Years later whenever I was giving aviation ground school or preflight instructional briefings, I tried to model my lecture technique upon hers.

The Air Cadet unit was many miles away, in Stafford. So, I would stay the night with my grandparents nearby and go directly to school from there next day. I was never actually able to attend an annual camp but summer weekends we would often gravitate to 632 Gliding School, which was then at Stoke on Trent (Meir) aerodrome. I would help out and, hopefully, get a flight. Thus it was that my first glider trip was in a Slingsby Cadet TX1, XA301 flown by one Flt/Lt Jackson, in November 1956. All I can remember about the flight was seeing some coal mine spoil heaps near Newcastle under Lyme and shivering with cold (and excitement?) for most of the time. We were airborne for, at the most, five minutes. It seemed like hours but from then on I was hooked. Things were so much more relaxed in those days, nobody even bothered to ask if I had a 'Blood Chit' authorising me to fly and signed by my parents.

Family holidays were always taken staying for two weeks with relatives in Warsash, near Southampton. In those days the area was teeming with civil and military aircraft from Eastleigh, Hamble, Lee on Solent, Portsmouth, Christchurch, Hurn and many others. But dad would rarely take me to local aerodromes and I spent a lot of time bored stiff on Southsea beach watching ships pass up and down the Solent. At least I could see the Aquila Airways flying boats and the Sunderlands at RAF Calshot, though never close enough to get their 'reggies' or take pictures. Unfortunately, Calshot closed for flying in about 1955, but not before the amazing sight of a US Navy Martin P5M Mariner flying boat had presented itself. We also witnessed some of the flights of the massive Saro Princess Flying Boat—even Dad took an interest in that!

So, one day in 1955, I set off on the bus and headed for Southampton Airport, at Eastleigh. I had a fine time there viewing the exotic contents of Captain Vivian Bellamy's hangar which contained a Supermarine S6B (the Schneider Trophy winner), a Gloster Sea Gladiator, the very last DH86B four-engined biplane airliner and Bellamy's own Spitfire. Unfortunately, I didn't have the bus fare home, so I caused a family sensation by arriving back at Warsash very late, having walked all the way from Eastleigh. The distance is ten miles, and it must have taken me hours.

In the summer of 1956, my mother was quite ill in

hospital, and we did not take our usual holiday. It was agreed that I could ride my new Raleigh Trent Tourist bike to Warsash. The bike had been financed by selling my electric train set, sweeping up in a printing works, as well as delivering bread and cakes from a baker's van on Saturdays, and various other part time enterprises. The plan was to make night stops at Youth Hostels near Banbury and Newbury. I set forth from Cannock at about five in the morning reaching Banbury at around eleven. I was a rather shy kid and had never actually been in a Youth Hostel before, so it seemed to make sense to press on to Newbury. It was only five in the afternoon when I made it to there, and again I elected to press on, arriving two days ahead of schedule to the amazement of the reception committee and with a rather sore backside. I'd clocked up 173 miles in about fourteen hours, having got lost twice. Wish I was still that fit!

One of the duties my part time job in the printing works entailed was firing up a gas-powered furnace, then taking all the used printing galleys and melting them down before pouring them into moulds to make lead ingots ready for the compositors to feed into the monotype machines. If a health and safety official found a fifteen-year-old kid doing that nowadays. he'd probably get the works manager put in gaol!

Some clean clothes arrived in the post. After a day of recuperation, I cycled to Shoreham, then to Eastleigh, Portsmouth and all the other aerodromes I could think of within a day's cycling. RNAS Lee on Solent had their air day the next weekend and, at last, I got my first flight in Avro Anson G-AHIB, flown by the late Captain Vivian Bellamy. The downside of this amazing expedition was having to get home again, but I did it in a day despite non-stop pouring rain. I arrived soaked but happy—at least cycling in the rain didn't make one so thirsty.

Next year I did the same again even finding time to stop at Oxford Airport for long enough to log and photograph its aerial occupants, and have a chat with the manager, former test pilot Peter Clifford. Many years later, working as a flying instructor, I was honoured to be able to renew his pilot licence for him.

Dave Whiteley, one of the typesetters in the printing works, was a great traditional jazz fan, and got me interested. I became a lifelong fan of bands like Chris Barber, Humphrey Lyttleton and Ken Colyer. Later my musical interests varied as widely as Mike Oldfield and Bach Organ music but I have always remained faithful to those British trad bands of the fifties and attended Chris Barber concerts until he passed away in 2021.

I became ambitious and, for thirty-five shillings, I bought an old Salvation Army cornet. Trying to teach myself was pretty hopeless and I couldn't afford proper lessons. In the end everyone bitched so much about the racket I was making that I could hardly ever practice, and the cornet was stashed away until my kid sister found it and ruined it. Probably just as well, for if I had ever been good enough to get into a band, I would not have cared much for the smoke-filled atmosphere in jazz clubs.

Many of the kids at school smoked. One of my chums was caught and 'interviewed' by The Bong. Before being given the whack, he was asked how many a week he smoked, to which he replied maybe two a month (lying git). The Bong recoiled in as much horror as if Ben had been on two packs a day. Fortunately for me, my dad gave me a drag on a cigarette when I was about eleven. I coughed so much it cured me of smoking for life.

Of course, we all fancied ourselves as 'Teddy Boys' in those days and admired anyone with drainpipe trousers, drape jackets, bootlace ties and duck's ass haircuts (crew at the front.) But at least 'Teds' were neat and tidy, unlike most of the hippies ten years later.

The fifties were the vintage years of the Goon Show, on the BBC Light programme. Most of us went around doing passable imitations of Eccles or Bluebottle, but the school teaching staff didn't see the funny side at all. I was not much use at memorising acts from Shakespeare, though I learned most of *Eskimo Nell* easily enough in the RAF.

I finally left school with only four Ordinary Level GCE

passes, and I got a lot of flak from my parents about that. Many years later. I discovered that was a slightly better than average result. Many boys left school with less, or none at all. Academically, it can't have been a very good school. Some kids did okay though, one in my year became a Navy helicopter pilot and another became an Oxford Classics professor, yet another made it to Chief Constable in another county. One girl even went back as a teacher and retired as head of the maths department. These achievements, however, were probably gained despite the school, rather than thanks to it! Anyway, on 4 April 1957, the notorious Defence White Paper had been announced by Duncan Sandys. All future military aircraft were to be unmanned rockets or missiles and the RAF was to be halved in size. I couldn't have picked a worse time to be hoping to get into the aviation industry.

Chapter 4 - The RAF

I found out about a job vacancy for a trainee mechanic in the Don Everall Aviation hangar at Pendeford Aerodrome. The manager, the late Eric Holden, gave me the job working on Austers, Tiger Moths, wooden Miles and Percival types and a DC3 Dakota, closely supervised, of course, by various ex-RAF types who actually did know what they were doing. That was where I had my very first driving lessons, on a six ton Crossley Type B ex-RAF fire engine or a Morris ambulance—when Eric had gone home! I also did a little bit of passenger flying in the Austers, and Dragon Rapides. In fact, the first flying I was ever paid for was sitting in the front passenger seat of a Dragon Rapide (G-AGDP) with a clipboard and filling out an air test form whilst Captain John Atkinson flew it. We lost an engine after twenty minutes and barely made it back to the field. My first in-flight emergency!

Health and safety would have had a great time inspecting that place. We were given no overalls and there was only cold water and cheap soap for hand washing. Most of the time it was freezing cold in the hangar. Messing about with all those chemicals including the old type red tautening dope, soon left one's hands in a pretty bad shape. Coupled with that, nobody appreciated the dangers of lead poisoning in those days. We would calmly wash our hands in high octane petrol then sit there eating our lunch with a patina of white lead oxide all over our mitts. Perhaps that's why I've always been a bit overweight. One of my chums was only fifteen, and he was hand propping Tiger Moths with the rest of us!

The pair of Pratt & Whitney R-1830 Twin Wasp engines on a Dakota feature fifty-six spark plugs. Removing, dismantling, cleaning and then reassembling and testing each one used to take most of a morning. Refitting them I was, of course, supervised by someone rather more competent. Refitting the engine cowls was always something of a drama for they never seemed to fit properly. One of us would lie sprawled along the top of the engine nacelle and squeeze the cowl flaps to the closed position. The hydraulic lines were the last thing to be reconnected so anyone on the ground had to stand well clear lest they get an eyeful of painful hydraulic fluid as it squirted out.

Airframe men have the unsavoury task of looking after the toilet systems. Simple enough in the Rapides and Dakotas, but there is a good story I heard many years later from a friend working on Boeing Jumbos. In their company they agreed that the airframe men would take it in turn to clear any toilet blockages, so they kept a roster of whose turn was next. One day it was the turn of a new guy. Equipped with a pair of very heavy-duty rubber gloves, he put a hand down the toilet to recover a Coke bottle. or whatever was jamming the valve. What he didn't know was that the pipes were connected and one of his mates had crept into the next-door toilet with another pair of gloves. As he groped around down the tube, suddenly, another hand grabbed his!

Pay was the princely sum of two pounds, fourteen shillings a week (£2.70p). There was a bit of overtime money, but out of that I had to give my mother thirty-five shillings (£1.75p), keep myself in bicycle tyres and light batteries, and cough up tuppence a cup for tea at break time out of the rest. That didn't leave a lot for living it up or dating girls. I think my parents saw that as an appropriate punishment for not doing better at school. Anyway, at the year's end, my father was foolish enough to quit his job, resign from the TA, and take on managing a hotel near Droitwich, Worcestershire. Eric Holden offered me a small rise if I would stay, and free accommodation in one of the old, cold, damp and draughty RAF billets on the aerodrome. But at that age I was pretty green and not really able to look after myself properly. Expecting to get conscripted into the Army (or worse still, the Navy) for National Service at any moment, I decided to escape that fate by volunteering for the RAF.

A very distant relative had been a Technical Branch Wing Commander at the end of the war, but since then had had no contact with aviation whatsoever. He advised me to sign up for a full career as an airframe/engine technician and, in retrospect, he was certainly correct. Unfortunately, in those days, there were not many opportunities for promotion to commissioned rank in engineering, and I certainly hoped to get a little further up the promotion ladder than maybe sergeant. Having heard all the usual horror stories about service life, I wasn't willing to sign up for more than five years at first— which seems a very long time when you are eighteen. Alas, the minimum engagement in which one could get airframes and engines was nine years. So, I became a camera mechanic, or, to use the formal title Photographer RS, which stood for Reconnaissance Servicing.

However, that was towards the end of National Service conscription and later it seemed that if I hadn't volunteered, I'd never have been 'called up.' On the other hand, it was a good time to join. To attract enough recruits and make up the numbers, the powers that be were making life a lot easier for the troops. Pay increases came through and most of the old army-style webbing was scrapped, leaving just the webbing belt to be adorned with immaculate blue grey 'Blanco'. Better quality uniforms were issued, and smart raincoats replaced the awful 'Groundsheet' things. Even 'Jankers' for minor miscreants, was replaced by the slightly gentler regime of 'Restrictions'.

After attestation and kitting out at RAF Cardington, we were whisked away to RAF Wilmslow for eight weeks of 'Boot Camp' or 'Square Bashing', officially known as Recruit Training. We were greeted on arrival by two

The inmates of block 246, RAF Wilmslow March 1958. Yes, it's me at the front.

corporals who were clearly a pair of prize bastards. In time, our view mellowed, and by the end of the course we even had a whip round for them.

Every 'Boot Camp' seemed to be at the top of 'Kitbag Hill', so arriving at the station we had to drag all out kit up to the camp, herded along by yelling and cursing drill instructors. It must have frightened the life out of some kids. Then we were herded into an open hangar and assembled in rows. Anyone who had been in the army, navy, cadet forces or Merchant Navy was ordered to take one step forward so that's what I did. Next thing I knew, being one of the tallest, I was appointed head of one of the barrack blocks with the title 'Senior Man'. It did me no good, it simply meant I was the first one in line for taking the shit when anyone else screwed up, since my bed was nearest the door.

The eight-week course was pretty tough on some of the guys especially those who had led a somewhat sheltered existence. Many suffered from homesickness, and some had a hard time keeping ahead of the game. We spent hour after hour on the drill square but, by the end of it, we were pretty slick. Ground combat training was a bit of a farce, with us yelling our heads off and galloping at a hanging sandbag, then stabbing it with a bayonet. At the end I was probably fitter than I've ever been, thanks to five mile runs in marching boots and other such joys.

It seemed that in every squad we had the 'Jock', the 'Taff', a 'Geordie' and 'Paddy' or two. Then there was always the sex maniac who never stopped going on about all his past conquests, and the 'Scouse' or the East Ender who fancied himself as a 'Hard Man', the gun freak—who eventually made a complete balls up of things when we got to the firing range—and the smarty-pants grammar school kid (probably me!). But it was a great levelling experience, and, in time, great team spirit and *esprit de corps* began to develop.

There was one national service guy on my squad who was quite a religious zealot. He got the lot of us into trouble for flatly refusing to fire a gun. He was going to be a medical orderly and in wartime would probably have registered as a conscientious objector. We tried to explain to him that guns were necessary in warfare not just to attack and kill, but to defend himself and his patients. After all, a medic has a duty to look after his cares, but no way could we convince him. In the end, the NCOs forced him to fire a rifle, but it reduced him to a dreadful state. The RAF Police took him away, so I don't know what became of him. He sure had plenty of guts and, in a way, I admired him, but I felt that his resistance was very foolish.

Many of us started smoking whilst we were there, maybe not because we wanted to but because of peer pressure. Every half hour or so the drill instructor would give us the 'Fall Out!' order, whereupon everyone would shuffle over to a corner of the parade ground and out would come the 'fags'. I lost count of the times I was offered one, but steadfastly resisted. My first priority was to save up for a motor bike, but sad to say, that's how many of us became addicted.

Apart from the drill we got minimal instruction in service procedures, air force law and so on. But the American VD film was a good laugh. Some fatherly looking US Army medical Colonel rabbiting on about syphilis, gonorrhoea, and soft chancre, with a brass band gaily playing joyful Sousa marches in the background. Then the station padre gave us a lecture on certain aspects of morality appertaining to relations with young ladies. Something of a wasted effort, for the rumour about bromide in the tea had spread around and we were so convinced about it that nobody reported achieving an erection for about the first three weeks. One Sunday afternoon, the catering officer strolled around the airmen's mess giving each of us a chocolate bar. The wrappers were of an unfamiliar design, so everyone thought it was laxative chocolate. Most of it went into dustbins but mine tasted okay and had no drastic effects.

I did score a good point at one lecture though. The instructor listed all the nine commands of the Royal Air Force at the time, except for the last one and asked if anyone knew what that was. Since nobody else seemed inclined to answer I put up my hand and said 'Ninety (Signals) Group, which has the status of a command'. Seems I was the only one of about a hundred who knew that. It came from my time in the Air Cadets, of course.

One drill instructor, a Corporal Jackson, could have made a career as a standup comic, with his lecture technique. His favourite party trick was to stand on one leg, with his uniform beret balanced on the heel of the other foot. Then he would kick it upwards and catch it on his head, usually the right way up. Anyway, they turned us into presentable airmen, ready for delivery to trade training at RAF Wellesbourne Mountford, near Stratford upon Avon, quite a pleasant spot to spend a summer. The wheel of fortune took me back there in 1981 as one of the founders of a flying club which still operates, and for years I kept my own Luscombe Silvaire aircraft there.

Not long after I left the regular RAF a theatre in Wolverhampton featured the Arnold Wesker play *Chips with everything* and seeing that certainly bought it all back to life. Judging from the howls of laughter at times, I reckoned half the theatregoers were ex RAF types. In the seventies, there was a TV series in a similar vein called *Get Some In!* on the same topic and that bought back a lot of memories.

RAF trade training was very well done, considering that many of our instructors were tradesmen with the rank of corporal or sergeant and not so very much older than ourselves. I guess it may have been easier for them than the job of schoolteachers, for we were all very willing to learn and very interested in the subject. Then, of course, there was the added incentive that course failure meant spending the remainder of the hitch we had signed up for in some lower grade job which would not be much fun at all. We were a pretty sharp course though, for everyone passed except for one unfortunate. He was the only one of us with previous experience working in photography and never stopped telling the instructors all about it. There are times when it's best to keep your mouth shut.

Some lessons in 'Trade Science' were conducted by officers of the Education Branch, one of whom was an extremely attractive WRAF officer, just a couple of years older than we were. She would strike a very seductive pose with the palms of her hands inside her uniform belt, her husky voice extolling the wonders of convex and concave lenses, so I guess we devoted rather too much mental effort on drooling over Miss Esme than listening to what she was telling us about. She must have been a good teacher though, since all but that one guy passed.

An end of course practical test was a lot of fun. There was a huge 'Dexion' structure built to represent an airframe in which all sorts of combinations of different aerial cameras could be fitted. Down the other end of the room we were poised and ready to go at the other side of a line painted on the linoleum. We had a selection of huge aerial cameras with all their drive motors, connecting cables, control units and so on. The examiner blew a whistle and we had to race over to the 'Sputnik', hauling all the kit with us and, working against his stopwatch, get it all fitted and working under a set time limit. Some of those cameras had forty-eight-inch focal length lenses and the films were nine inches wide, so they were pretty heavy items to lift and manoeuvre in a confined space.

I was to be posted to Cyprus. After a couple of weeks of embarkation leave, spent touring around civil aerodromes on my 150cc James motor bike, we reported back to Wellesbourne and, after the usual 'clearing' procedures, entrained for RAF Innsworth, near Gloucester. Our left arms were ruthlessly jabbed with multiple inoculations, and we were 'kitted out' with all the tropical gear we would need. Then it was another train ride to Southampton and truck to the dockside where our 'cruise liner', HMT *Devonshire*, awaited our pleasure. The voyage took two weeks and was not something I would recommend. At least we had fold-down bunks and not hammocks, but it was quite crowded and the food was utterly disgusting. After a couple of revolting dinners, I resorted to living on NAAFI slab cake and tea for two weeks.

The ship was operated by the Bibby shipping line, which also owned Skyways, one of the major charter airlines of those days. All other troopships were for the army, but this was the only one with a permanent RAF presence, presided over by the 'Ship's Warrant Officer' who spent most of his time devising shitty jobs for the 'passengers' to do. Fortunately for us, he gave most of the shitty jobs to the army majority, so mostly we pulled 'Man Overboard' watch duties. It was the first long sea trip for most of us, but the rough crossing of the Bay of Biscay didn't give me any trouble, so my sea legs seemed to have developed okay. Maybe those canal trips in the Sea Scouts did do some good.

One morning I *was* seasick though. We'd crossed the Bay of Biscay without any ill effects until one day when I was on 'Man Overboard' watch at the port side stern position. We were cruising along in glassy calm water at the time, and I reckon what caused it was the revolting stink that came up from the engine room ventilators, or perhaps the pong from the stuff the Lascar seamen were eating for breakfast with their fingers.

Most of the troops were Kings Own Yorkshire Light Infantry. It was just after the *Last of the Mohicans* had been a popular TV show, so most of them were sporting 'Mohican' haircuts. As soon as we reached warmer climes, they all took to sunbathing whenever they could and most got badly burned. Red sunburn coupled with all their blue tattoos and Mohican hairstyles made them look terrifying enough to frighten the enemy to death— I figured they probably wouldn't need guns.

Our first stopover was in Gibraltar, where we were permitted ashore and spent the day wandering around looking at the shops and cafés—we'd hardly got any money anyway. At Malta, I didn't even bother to go ashore—the inspection on leaving and reboarding was too much hassle. If only I had known it at the time, I could have visited the grave of a distant relative after whom my father and I were named. He was fatally injured at Gallipoli in 1915 and buried in Valetta, but I never found out about him until 2016. Anyway, at least some of the troops disembarked there, so we had a bit more space from then on.

In Cyprus I was not expecting to do much aircraft photography, though the variety of transient military aeroplanes was quite interesting. First some admin type made a mistake, and I was told to report to RAF Nicosia. So, after a most unpleasant few hours in the back of a three ton truck, I had to head all the way back to RAF Akrotiri. We all got pretty thirsty with nothing to drink so nothing has ever tasted better than the cold-water fountain in the airmen's mess when we arrived. Initially, I joined Canberra PR7 equipped 13 (PR) Squadron, which shared a dispersal with the Station Flight, who looked after visiting aircraft.

Initial accommodation was in tents, although that's not so bad as it sounds. The ridge tents were on concrete bases with breeze block walls about three feet high. So, with a slight sea breeze they were cool enough in summer. Then some ace medical officer decided they were too cold in winter, so everyone was moved into the overcrowded rooms. The 'Caywood' bungalows that served as barrack blocks were designed to take six men to a room. With the steel frame beds stacked up double, we were packed sometimes as many as fourteen to a room. Enter a flu bug, and it ravaged through the camp like lightning.

Another problem was the bed bugs. All the barrack furniture had been shipped over from the Canal Zone when the RAF pulled out of Egypt, so it was infested with the damn things. They were not hazardous but their bites itched and if you killed one it stank. Now and then we'd get completely fed up with them and spend an afternoon clearing every last item out of the room. Then we'd wash the floor and check over every item of furniture before it was hauled back in. Using a paraffin blowlamp, we'd burn the little beasts in their hideaways. Then the beds would be put back with each leg in a nine inch film tin half filled with water and a drop of jet fuel floating on it. Even the seams of the mattresses got the blowlamp treatment.

Photography of aeroplanes was not actually banned but was not encouraged, although I do still have a few good shots of Canberras, Meteors, Pembrokes, Chipmunks and various exotic visitors such as the Savoia Marchetti SM82 of the Lebanese Air Force. Happily, in 2006, I was able to respond to an appeal by the station museum and send them a number of suitable images.

The squadron had many commitments all over the Middle East and much of Africa so, once the EOKA emergency had died down, we flew to Nairobi (Eastleigh) aerodrome, in 70 Squadron Hastings via El Adem, Khartoum and Aden. Thus had the RAF kindly provided a layabout aeroplane spotter with free afternoons hanging around Nairobi West (Wilson) Airport. Khartoum airport was something of an experience. We'd thought Nicosia was pretty hot when we left, but Khartoum was quite something else. Exiting the Hastings was akin to opening an oven door to inspect the Sunday joint.

Waiters clad in white gown-like garments glided around the restaurant, serving us with cups of strong Arab coffee. To our great delight we didn't have to pay for it because the local BOAC station manager kindly coughed up. Just as well, because most of us had hardly any money anyway.

Late night arrival in Aden was quite an experience too,

although the place was somewhat overcrowded, and we had to sleep in the open on the verandah of the transit block. There were some strange uniforms in the airport transit lounge too. Seconded army officers in Arab headdresses made us half expect to see Lawrence of Arabia sweep into the room at any moment.

Years later, I heard a tale of an Eagle Aviation crew that were staying there at about that time, in the Rock Hotel. They'd been out somewhere else for 'a few drinks' and staggered back into reception from their taxi. Through his alcoholic haze, the first officer somehow focussed his eyes upon a large rug, which appeared to be impregnated with years of aircrew puke etcetera. 'Whaddya want for the rug?', he says to the night manager. Somewhat perplexed, the night manager couldn't believe his ears. eventually, in the hope the crew will just clear off to bed, the night manager tells him, 'Give the house boys a couple of quid and you can have the darned thing'.

So, under much protest, but really treating the whole affair as a bit of a laugh, the crew drag the filthy thing up to Jim's room. Next day it is dragged out to the taxi, and dumped in the forward hold of their aeroplane, though by that time the rest of the crew would have preferred him to chuck it into a storm drain.

On arrival at Heathrow, some days later, the customs man takes one look at it, holds his nose, and announces 'You can have it, just get the bloody thing away from me'. Jim took it to a professional carpet cleaner near his home in Maidenhead and then on to a specialist for valuation. It was a genuine antique Esfahan carpet, maybe a couple of hundred years old, and was worth hundreds, possibly thousands. The kind of carpet intended more for hanging on a wall rather than laying on a floor. Jim sure knew his antiques. Even when he was half pissed!

Our detachment commander was trying to get us seats on a civil airline, but after a few days we got a lift in another Hastings and arriving in Nairobi. I guess, seeing Kenya for the first time, we all thought we'd died and gone to Heaven. RAF Eastleigh was probably the most beautiful station in the entire RAF, with flower beds and shady Jacaranda trees here and there. Apart from us, there was only one small light transport squadron based there at the time, and a unit doing major checks on Shackleton bombers from the Aden squadron. A few men of a signals unit and a few army blokes meant that the base wasn't overcrowded like Akrotiri or Aden. Perhaps best of all, the food in the airmen's mess was significantly better than what we'd been putting up with at Akrotiri.

There was no official 'apartheid' in Kenya, though double deck buses had hard, second class, seats upstairs, which whites were not expected to use. There was no segregation on single deck buses, so one day I headed out to the new international airport at Embakasi. I am afraid I was caused more than a little embarrassment when a black lady sat down, right next to me, and promptly began breast-feeding her 'mtoto'.

I photographed a few East African Airways Dakotas, Hunting Clan Viscount G-ANRS and saw my first BOAC Britannia 312. A few years later I was doing flight plans for that very Viscount and navigating that very Britannia with British Eagle.

The corporal in charge of the base photo section owned a beaten-up old station wagon and offered to take us out to the Athi River game park one Saturday. We spent some enjoyable hours watching big game and eventually parked a few yards from a pride of lions. Night falls quickly in Kenya, so when it was time to move on the wretched truck wouldn't start. There was only one thing for it ... get out and push. I can assure you that doing so with a bunch of huge moggies snoozing fifty feet away puts a very great strain on the nerves—despite our corporal friend's assurances that it was safe because they'd already eaten. All right for him, he was in the drive cab! Thankfully the push start worked.

On Sundays, the whole bunch of us would go to lunch at 'Hoppy's Ngong Inn', at Karen. Hoppy Marshall had been clerk to the District Commissioner when the Mau Mau rebellion began, but the DC didn't have a hangman. With business being brisk, and a fee of £25 a time, Hop-

Hoppy Marshall, the Kenya Government Hangman. His belt-mounted gun, which was always cocked, is just out of vew.

py had volunteered for the job. His pub and restaurant served up some pretty good tucker and lots of society ladies from the city pitched up to hear him cussing and blinding, since his vocabulary of invective was quite spectacular. His waiters were rumoured to be ex-convicts and he always carried a gun. Out of professional interest, one of our armourers reached to have a look at it but Hoppy swiped his hand away; the gun was cocked with one 'up the spout'. He was called Hoppy, by the way, because the weapon once went off and shot him in the foot. He needed the gun because he'd had numerous warnings, including a dead cat nailed to his front door. When I went back there after Kenya Independence in 1966, he had left for the relative safety of South Africa.

We flew back in a Beverley as far as Aden, where we had to wait for a Hastings to give us a lift home. We were stuck in Aden for about a week, so didn't miss the opportunity to stock up with duty free cameras, binoculars and transistor radios etc. We were a crafty lot though, and not averse to a little, well ... perhaps *more* than just a little discreet smuggling. We certainly didn't take the Cypriot customs very seriously, after all, their declaration forms asked us to solemnly swear we were not importing any 'meat

*The corporal assured us they were safe because they'd just eaten. i
wished I could share his confidence. Nairobi Game Park late 1958.*

sandwiches'. Cigarette rolling machines, however, were
taken very seriously—I believe hashish grows wild in some
parts of the island.

So it was that among the tons of huge survey film tins
that were flown back in the flare bays of the Canberras,
were other identical tins marked 'Survey Film – TOP
SECRET'. These were all whisked away under police guard
to the photo section, where even police were not allowed
to enter without good reason, and Cyprus Customs had
no chance. Then the unpacking of transistor radios, Leica
cameras and expensive binoculars, etcetera, began. Of
course, we eventually had to get all that loot back home
to the UK, but when I did come back I found Stansted
customs very generous to returning servicemen, and I'd
only acquired one small 35mm camera.

The EOKA emergency was very badly organised. It
had been going on for about five years by the time we
hit town, yet the only fortification around most of the
aircraft dispersals was a couple of rolls of barbed wire.
That changed one night when I was patrolling around our
dispersal at two in the morning. I heard this loud explosion
about half a mile away, and then the hue and cry really got

going. Land Rovers full of RAF Regiment blokes charging frantically around the airfield, then a coach load of crew arrived to fire up a Shackleton which eventually staggered into the air and spent the next two hours circling over the base dropping illumination flares. A bomb had gone off in the tailpipe of a Canberra bomber belonging to a visiting squadron, parked on the ASF site.

Some weeks before I arrived an entire hangar was burnt down. Some say it was an accidental fire, others claim it was an EOKA attack. Either way it resulted in the destruction of three Canberra bombers. There had been a few other desultory EOKA attacks, a GCA trailer on the airfield was damaged the night I arrived, and before that the half-built NAAFI had been blown up. This time though, the three Ground Defence RAF Regiment squadrons really had to get their act together. Miles and miles of barbed wire appeared and eight foot high double fences were built around all the dispersals. Between each were stacked with more rolls of barbed, then another double fence about twenty feet outside. The 'killing area' between the two fences had strands of barbed wire zigzagging between posts about a foot high acting as very nasty tripwires. Each dispersal had one or two 'Goon Towers' about twenty feet high, with plenty of sandbags around the covered platform, making the place look a lot like Stalag Luft Akrotiri.

Annual refresher ground combat training was supervised by RAF Regiment NCOs at the firing range on the cliffs. I was always a lousy shot with a Lee-Enfield but quite enjoyed blasting away with a nine-millimetre Sten gun. Firing the Bren was great fun and, over the years, most of a large rock a few hundred yards out to sea was chipped away to nothing. Number two on a Bren wasn't much fun, ready to change barrels with one's head about a foot from the muzzle. The noise was excruciating and really made your ears hurt badly. No ear protectors in those days.

The revolver technique was a farce. Using the left hand to steady the right wrist as in American gangster films was not allowed. Instead, you had to hold it at arm's length which clearly reduced accuracy. It's amazing how if you

Front line of the nation's defences. RAF Akrotiri 1960. NDW third from right.

take a careful aim, you often miss, but if you just blast away, you often hit the target. I couldn't help admiring armament sergeants who could keep a tin can bouncing fifty feet away with continuous shots. Something to do with the subconscious mind over-riding the conscious.

The RAF had wanted to raise levies somewhere else in the Empire and ship them over to Cyprus to take on some of the guard duties. Typically, though, the ever mean and parsimonious British Treasury, as always concerned more with money than the lives of servicemen, said the airman could do all that in their spare time. The result was that everyone did a full day's work with a twenty-four-hour guard duty on top of it, every three or four days. Add a few parades and 'bull nights' as well as keeping one's own kit clean and tidy, and the inevitable result was that everyone completely knackered, and it was reflected in their work. Maybe that was one of the reasons so many Canberras were damaged in belly landings because the wheels wouldn't lock down.

The first one of those I witnessed was a masterly piece of flying by a Flight Lieutenant Norris. The left main-wheel of a Canberra trainer failed to lock down so he greased it down on the right wheel and nose-wheel, keeping the left wing off the deck with right aileron as long as he could so that when the tip tank hit, he was moving so slowly that the aeroplane swung about thirty or forty degrees to the left and stopped. Rather like watching a sailplane land. I believe he got the Air Force Cross for that. If not, then he certainly deserved it.

Our Flight Lieutenant Peter Clayton was the next victim of an uncooperative undercarriage. We watched him pulling 'maximum G' turns above the airfield and burning off fuel while the fire trucks go into position. At that stage, the runway 'foam carpet' technique had yet to be invented, so, with most of the fuel burnt off, he greased it down on to the dirt crash strip. The spark it generated when the bottom of the fuselage impacted was about three times the length of the aircraft. Soon as it stopped the cockpit door burst open and I'd never seen anyone move so fast. The main fuel tank was in the fuselage belly and that could have gone up any second.

The third one I saw was not long after that, pulled off by Flying Officer Neil Williams with navigator Jack Stratton. Neil later qualified as a test pilot and became an international aerobatic champion. He was an exceptional pilot who once landed an aeroplane after the wing spar broken, making his approach upside down and rolling over just before he landed. Witnesses claimed he was so low that the tip of the good wing actually scraped the runway as he rolled. He got a Queen's Commendation for that but a few years later was tragically killed flying an old Heinkel bomber from Spain to a UK museum.

Once I was detailed for truck escort duty. I asked for a Sten gun, which would have been easier to manipulate inside the truck cabin, but no, it had to be a .303 Lee Enfield rifle. Five rounds of ammo were all I was given. With an older LAC who wasn't armed at all, we set off for Nicosia, about eighty miles away. His instructions were to ignore all speed limits and drive flat out through all towns and villages. Unofficially we were also told that if anyone looked like they were throwing anything at us, shoot them and keep going. If I killed a stone throwing child, I would almost certainly get fifty-six days in the military prison at Waynes Keep but that would be better than the risk of ending up next door, in Waynes Keep Military Cemetery.

I suppose we were too young to appreciate the dire danger we were in, that was about the same time that half a dozen RAF airmen were killed by a bomb hidden in the Malcolm Club jukebox. If we had been ambushed or hit a roadblock, we would have had no chance and surely would have been killed or captured before there was any chance of backup arriving. We had no phone or radio contact anyway.

One guy got himself into a bit of trouble for sleeping on guard duty. The taxiway into each dispersal was blocked by a single roll of barbed wire that had to be pulled out of the way by the lad in the tiny guard hut. Mid-morning a Canberra taxied up to the wire, but nobody emerged to remove it. Revving the engines had no effect so eventually the exit door opened and out climbed an officer who had been flying supernumerary on the jump seat. Imagine the shock horror when the airman was shaken awake by an officer with Air Vice Marshal bars on his overalls.

Service always had its lighter, often hilarious moments though. One chum was son of a very senior civil servant working in the Air Ministry, with a very high 'Equivalent rank'. On a 'charge' for some petty misdemeanours, Dave was duly marched into the office of the Squadron Leader, along with another mate acting as 'Airman's Friend', as per the required procedure. Dave had the usual smirk on his face, so the boss made some remark like 'I suppose your you're going to tell me your father is a Wing Commander?', to which Dave replied, 'No Sir, actually he's an Air Commodore'.

At that point everyone else in the room collapsed into fits of giggles and the 'Airman's Friend' actually wet himself. So the boss threw the lot of them out. The hearing was never re-convened though. His father did actually have the 'equivalent rank' of Air commodore, as a senior civil servant statistician in the Air Ministry.

The squadron operated from many out stations such as Bahrein, Sharjah, Masirah, Salalah etc, where there were no photographic staff stationed. So, two camera mechanics were always put through the decompression chamber and cleared to fly as supernumerary crew on the jump seat of the Canberras. The two who were current in early 1959 were due for return to the UK. One, George Henderson, soon took a job with Skyfotos at Lympne but tragically lost his life not much later when a Piper Tri Pacer was lost at sea after doing some ship photography. Not many of the airmen were very interested in flying and there was little competition for these seemingly attractive slots. The word was out that I was interested in applying for an aircrew course, so my name was put forward and accepted. This was not without some risk. One of our aircraft had been shot down over Syria a year or so beforehand and one of the crew, Flying Officer (Navigator) Urquhart-Pullen killed. Typical of the treasury-hamstrung RAF, there was no flying pay, promotion or aircrew status. Shades of Cpl Alan Fox who in World War Two flew eighty-five missions in Mitchell photo recce aircraft over Japanese occupied Burma and was not even given the rank of sergeant to protect him event of capture. At least he was one of the very few RAF ground crew to be awarded the DFM, as were camera mechanics LAC Hadden and Cpl Shirley, in Malta., who had provided much of the intelligence photography upon which the Royal Navy attack on Taranto was based.

In addition to the regular mapping survey and reconnaissance flying, 13 Squadron was sometimes tasked with other jobs. An example was the discovery of 'The Lady Be Good', an American B24 Liberator bomber in the Libyan desert. Found by an oil exploration crew, it had lain there since it ran out of fuel and crashed one night in 1943, with no crew surviving. A Canberra was sent there to take low altitude oblique photos of the wreck.

Another was the crash of an Avro Tudor freighter full of highly classified rocket parts on its way to Woomera, Australia. That hit Mount Suphan close to the Russian border, killing all on board. It was probably lured off its track by the Russians manipulating radio beacons. The wreckage had to be photographed from low level too. Canberras were sometimes used to look for EOKA roadblocks, but that was so uneconomic that the only operational Chipmunk squadron ever formed took that task over. All that flying was, of course, very interesting for our aircrew, but the high level of security prevented any of us keeping prints as souvenirs.

As it happened, I never did get to fly in a Canberra as I made the stupid mistake of lacing my boots up too tightly before going on a wing parade. Circulation in my legs must have been affected and I passed out in the heat. Squadron leader Charles Crichton decided that I was not fit enough to fly and wouldn't listen to any excuses, so I was removed from the list. There was one way of getting some flying in though. Two maritime reconnaissance Shackleton bombers shared our dispersal, and if one had the spare time, riding in the back of those was quite encouraged. Usually, they flew trips of about eight hours, circling the whole island and investigating any small boat within about fifty miles of the coast. If a boat looked like an arms smuggler, a naval vessel would be vectored towards it. Unfortunately, that ceased soon after the end of the emergency, when the 'Shacks' returned to their parent unit in Malta.

Soon after that for, some reason, I got on the wrong side of a certain odious Flight Sergeant. He had put me on a charge over some petty matter, but Sqdn/Ldr Crichton threw it out and Flt/Sgt McClean got his arse reamed for wasting the Boss's time. He got rid of me when another photographer, mis-employed on a bomber squadron, applied for an exchange posting to thirteen, so he could do some real photographic work. I was sent to 249 Squadron, which was a bigger unit but just had one bomb strike recording camera per aeroplane. These were almost never used so all I had to do was keep them serviceable. The rest of the time I helped out in stores, the admin office, did amendments for the navigation office or minded the telephone.

Each morning we jolly crowd travelled to work in the

back of a Bedford three tonner, and it wasn't advisable to miss it. An unavoidable hazard lay on the route in the form of an RAF Police hut, populated by a number of mean-spirited creeps we called 'Snoops'. One especially obnoxious Welshman delighted in putting airmen on charges for tiny errors in dress, ignoring the fact that minutes later most of us would either been in grimy overalls or nothing more than a pair of shorts and flip flops. However, there was one short cut over the bundu, although that entailed climbing over rolls of barbed wire. Aged nineteen I was probably fitter than I've ever been and soon found that with a suitable tailwind I could traverse the whole lot. Alas, a rather more rotund colleague attempted the same and landed squarely in the barbed wire coils resulting in a painful visit to the medical officer. I was told the MO subsequently wrote a paper on perforation of the testicles, which received much approbation in medical circles and may even have led to his elevation to the peerage.

If I had had any sense, I would have borrowed training notes from the engine and airframe fitters and applied to take the trade tests to re-muster, since it was permissible to do that without extending one's service. Instead, I slapped in my application for aircrew training as a navigator and bought a pile of textbooks on air navigation to study.

The first reaction to this was a visit to my tent from my flight commander, the then Flt/Lt Terry Gill, with whom I still keep in touch. Terry tried to persuade me to alter my application to pilot, but I was reluctant and maybe a little too short on confidence to think I could make it. Then the adjutant, navigator Flt/Lt Chris Podger—a former monk—came to see me and find out how much I knew about navigation. He was suitably impressed, and both okayed my application. There then followed a progression of interviews with Sqdn/Ldr Douglas Barfoot, our Commanding officer, Station Navigation Officer Sqdn/Ldr Harry Scott, who wore the pre 1941 Observer brevet, and finally a short formal interview with the Station Commander, Group Captain Andrew Humphrey. He had commanded some of the 'Aries' navigation research flights over the North Pole and later became a Marshal of the Royal Air force and Chief of the Air Staff, tragically dying in post at an early age.

Somehow, I managed to favourably impress them all, but the blow fell when I was ordered to report for an aircrew medical. I knew my right eye might be a problem since I had astigmatism in both eyes and had worn glasses from age six onwards although, from the age eighteen, I found I did not need to use them other than for reading in poor light.

Unless a candidate had professional qualifications, there were few commission opportunities other than as aircrew. I was offered the RAF Regiment and, oddly enough, did seriously think about it. I had no real interest in guns, and didn't object to ground combat training, assault courses, parade drill and all the other military stuff—I had grown up with it after all. What put me off was football. There were three Regiment squadrons on the base and all they ever seemed to do was play football. I had been surrounded from birth, at home and through school by messianic football zealots and was totally sick of it. Still am. I sometimes enjoy watching rugger though I know almost nothing about it. It's just the fact that nobody ever gets killed that fascinates me!

Mind you, even if I had achieved aircrew rank and status I doubt if I would have progressed very far in promotion since some interest and participation in sport is usually a requirement, and that's never been my strong point. In later years the technical branch expanded a lot and created many opportunities for airmen and NCOs in engineering trades to progress, even as high as Wing Commander and beyond, but of that I learned too late.

Eventually, my boss sensibly concluded that my presence was a waste of taxpayers' money and got me a posting back to 13 Squadron. They didn't even need me on the flight line, so I was sent to the main photo section to help out, though I wasn't qualified in and knew precious little about that side of the business. So, I spent most of my final few months sitting in the guard hut signing visitors in and out.

It even gave me the time to read Tolstoy's *War and Peace*, both volumes.

I did have to do some productive work once during an exercise, and that nearly had an unfortunate result. The processing machines were working flat out, and I was told to help the crew on one of those. The final bath the film went through was a forty percent solution of methylated spirits and water, so that when the film entered the drying drum, it would dry just that bit quicker. As final man on the crew I was right by the meths tank in the sweltering heat, and I must have breathed in too much meths vapour. The boss then decided to hold the pay parade and a relief man arrived so I could collect some cash. I could hardly stand up but staggered over to the boss's office. He was a lovely chap though and realised soon enough what was up. He ignored my sloppy salute and ordered me to sit down and get some fresh air. My word, what a headache I had for a couple of hours.

That was probably the reason that despite many years of readily available, cheap, duty-free booze, I've never been drunk. As soon as I have had enough to get a certain feeling, I stop drinking right there and then. I've even been known to tip a double scotch into a flowerpot. Fortunately, I've never liked beer or wine, which is just as well since my family tend to be overweight.

One spectacular event was the massed arrival of four squadrons of Hastings transports. They were lined up one morning along the parallel taxiway and one by one staggered into the air. Climbing away into the distance they all swung round and came back over the base in formation. As we watched the parachutes opened and hundreds of soldiers floated down to the nearby salt lake drop zone. With more than fifty aircraft—each loaded with maybe fifty troops—I guess we saw nearly two divisions coming down. It gave us some idea of what the start of the Arnhem battle had looked like.

Each Christmas it was the custom to empty out a room in every barrack block and turn it into a bar, each one decorated in a different style. We decided the theme of ours would be the Goon Show and we did some pretty impressive artwork. The trouble started when 'Groupie' came to decide which bar won the competition, he'd no idea what ours was all about. Made no difference anyway, it was always the one decorated to look like 'Ye Olde Englyshe Pubbe' that won!

For a couple of hours one evening all the officers and their ladies came along for drinks, so everyone was on their best behaviour and decorum reigned. Not so on subsequent evenings when raucous singing of such bawdy epics as *Eskimo Nell* or *The Elephant and the Kangaroo* could be heard from afar.

With about four thousand blokes aged, on average, about twenty, and only about a dozen WRAFs on the base, you would assume it to have been a testosterone time bomb. A few of the guys used to sneak off to the off-limits 'Zigzag Street', running the gauntlet of RAF Police patrols for their assignations with local hookers, but I think the VD films had put most of us off that style of diversion. Nor do I remember any cases of homosexual activity despite so many having to live in such close proximity. If anyone displayed the slightest signs of effeminacy, the 'piss taking' was quite merciless—homophobia reigned supreme.

Sometimes, a spell of boredom may be a good thing because it encourages initiative to develop. Just for a laugh, someone had the bright idea of making an imitation gun, so the idea sort of caught on and using a bulk 35mm film tin I made myself a passable replica Tommy Gun. Soon, pretty well everybody had one. Then someone got the huff about it and started to complain, but in the end those 'set in power above us' decided we were all nuts anyway so they may as well leave us to it.

Someone had the bright idea that an all-ranks beach party would be good for morale, so three or four truckloads descended upon the beach at Aphrodite's Rock on the coast to the west. With the party and barbecue in full swing, some passing staff officer from Air HQ spotted it all from the roadway, so asked his driver to stop so he could come and say hello to us. That was his big mistake for he was soon grabbed by a bunch of drunks who wanted to throw him into the sea. Eventually a compromise was reached so he got launched just in his hat and underpants. He took it all with good grace and after a few drinks and a burger with us went off on his way. A good laugh at the time maybe, but saltwater damage to his hat could have been serious, and Group Captain uniform hats don't come cheap. The day ended with about five of us climbing to the top of Aphrodite's rock, then standing in line pissing into the sea.

Towards the end of my overseas tour, we were once more allowed out unarmed and in civilian clothes. Late one afternoon myself and friend had just wandered out of a bookshop in Limassol when we heard a gunshot. All the street hubbub stopped instantly so we figured we had better make ourselves scarce. From nowhere suddenly appeared a uniformed RAF Police Corporal who, without saying a word, opened a side door in a wall between two shops and shoved the pair of us through. That led to another street where we hopped the first bus we could get back to base. Later I heard it was not any resumption of EOKA activity, but a vendetta going on between two Cypriot families. That was about the only time I'd ever been pleased to see an RAF policeman.

At last came the day when we were due to come home, and the bus sped out past the guardroom with the lot of us singing an insulting ditty about the RAF Police—as loud as we could—as was the custom. At Nicosia transit billet I found myself in a bed next to a soldier who had overfilled a soft sided suitcase so badly it had burst. He hadn't a clue what to do about it so a few of us took pity on him, nipped outside, and stole the clothesline from the RAF Police billet. With that, we lashed his suitcase back together. Next morning, we cheerfully boarded a Bristol Britannia of Freddy Laker's Air Charter company and six hours later we were shivering on the ramp at Stansted. By then I guess the police were still scratching their heads in wonder as to why anyone would wish to nick their clothesline.

In later years I flew more than half the Bristol Britannias that were ever built, but I never got to fly in G-ANCE again.

Chapter 5 - Home Again

Returning from Akrotiri early in 1961 I was posted to RAF Kinloss, Morayshire, but first had a few weeks disembarkation leave. I rejoined the Aero Club back at Wolverhampton and began flying lessons with Captain Louis Wood and his assistant, Chris Lloyd. Two Tiger Moths were available for £3/10/0d (£3.50p) per hour but one had to dress up to fly in those, especially as it was still February. So, I paid the extra 2/6d an hour to fly the Austers, in retrospect, a somewhat foolish decision. In fact, I was stupid not to have at least taken a half hour trial lesson on a Tiger Moth whilst in Kenya. VP-KJV or VP-KCZ would have been a much nicer first line in my pilot logbook.

I did about ten hours and was almost at solo standard but my posting to Kinloss meant that the nearest flying club was at Perth, well over a hundred miles to the south. So, I had a few more lessons at long intervals whilst on leave, but gradually lost interest.

At Kinloss I scrounged a few more trips in the back of Shackletons, mainly on the pretence of trying to photograph ships, many of which were suspected of being Russian spy trawlers. There was a big security clamp on all of our shipping photographs, especially at the time of the Cuban Missile Crisis, but imagine our amusement when perfect pictures of the same ships appeared in the national papers. They had rented light aeroplanes and caught the Russian freighters as they passed through the straights of Dover. The weather being clearer down there made their shots much better than our 'Top Secret' ones. Anyway, at least it gave me some freebie flying time and gave me some idea what navigating a large aeroplane was all about.

A tragic accident happened one winter morning. I was walking back to our office from one of the aircraft, when I met the duty technical officer as I entered the hangar. I saluted and we exchanged the usual courtesies then off he went into the darkness. Seconds later he was dead. The first thing I knew about it was a few minutes later as I came out of our office again and was accosted by an airman who was near hysterical. He had a strong Geordie accent, so it took a while for me to figure out what he was babbling about. Anyway, a Shackleton had started up and because the ground crew were having trouble moving the GPU (ground power unit) away, the officer had gone to help but slipped on the ice, fell into a propeller and was killed instantly.

Our Warrant Officer later had the unsavoury task of photographing the remains and the damage to the propeller. I was absolutely amazed to see that about the last two feet of the huge aluminium blade had been bent about one foot at the tip. A full military funeral took place a few days later followed by a Court of Inquiry. I didn't think there was any point letting on the fact I was probably the last person to speak to the poor chap, so I managed to avoid getting involved in either.

Kinloss wasn't having a lot of luck in those days. Just before I arrived the station commander, Group Captain Avent, had suddenly died and a major disaster struck soon after I left. An airman I had known slightly decided he needed a cigarette, so took a stroll out along the taxiway from the torpedo section. Seconds after he settled down in the grass and lit up, there was a loud 'Wooomph' and a blast of air knocked him over. The torpedo section had blown up, instantly killing all within. Smoking may be fatal but, on that day, it saved one life.

Another interesting event was the discovery of some mysterious wreckage at about 2,000 feet up a hillside near the old RAF station at Evanton. The wreckage was collected by the Kinloss Mountain Rescue Team and dumped in the

I had my first flying leson in this Auster in early 1961. I flew her again in Australia 47 years later.

photo section studio whereupon an immediate security blanket descended. RAF Police on the front door, and all that. A nameless flight lieutenant appeared from London and gave us the usual cautionary reminder of the Official Secrets Act— Tower of London and all that guff. Then he had us photograph all the junk from every angle before it was crated up and shipped away to who knows where.

We figured that it was some kind of Russian weather balloon, although all the details didn't quite fit. Then, many years later, I was flying with an airline Captain who had been a RAF pilot in the fifties. At one stage of his career he had been asked to report to the Air Ministry to be interviewed for a special job. This turned out to be acting as the RAF representative on a USAF unit that was launching high flying photo-reconnaissance balloons from RAF Evanton. The plan was for them to drift over Russia at about a hundred thousand feet, way up above where any flak gun or jet fighter could get them. Then, as they flew out over the Pacific, they were to be retrieved by specially adapted aircraft which sent radio signals causing them to drop their payloads. Not longer after that, the whole story came out with the publication of the book *The Moby Dick Project*, by Cliff Peebles.

There were, in my mind, a few lighter moments. Weeks in advance of the event, we were notified the Queen would be making an official visit. Working parties frantically tidied the place up whilst the station ground defence officer got a few dozen unfortunates back up to parade drill standards. On the great day all private cars were removed and hidden away somewhere, and all personnel were ordered to stay out of sight. Naturally, I was keen to at least catch a glimpse of Her Majesty and, as her limousine swept by, I was able to do so by standing on a toilet seat and looking through a lavatory window. I therefore lay claim to have been one of the few, maybe the only, person who has photographed Her Majesty through a lavatory window.

Sometime later I heard of the start of a General Duties (Ground) Branch being formed which would embrace Air Traffic Controllers, Fighter Controllers, and Photo Interpreters. Until then most of these jobs in peacetime had been done by aircrew between or after flying. There were opportunities for Photographic Officers, although these were very few and were mostly taken by former aircrew. My eyesight precluded Air Traffic Control and I was not much interested in working as a Fighter Controller in some underground radar bunker, so it was that I was accepted for training as a photographic interpreter. Somehow, they decided to offset my defective eyesight against my acquired photographic knowledge, such as it was.

Then, for some reason I could never quite figure out, the RAF sent me on a two-week meteorology course at Birmingham University. It was conducted by a senior forecaster from the Meteorological Office and held at the Geography Department Field Station on the old RAF Station at Montford Bridge in Shropshire. It was an incredibly interesting two weeks, learning all about observation instruments and using a theodolite to track a met balloon. That was, of course, long before the days of pocket calculators so it often took us a whole evening to translate our readings into wind speed and direction, just using trigonometrical tables and slide rules. Later, when I was studying for a flight navigator licence, the experience turned out to have been very useful.

Accepting the promotion, however, would have meant signing up for another nine years at least, and the age ceiling for the maximum likely rank of Flight Lieutenant— if I could stay longer—was about forty. I could not envisage much of a future career for a forty-year-old former photo interpreter so, when my five years was up, I decided to leave the service, and try my luck in civil aviation. If I'd taken that option maybe, I'd have ended up in MI5!

Fortunately for me though, the Navigation Leader of the Maritime Operations Training Unit, Flt/Lt Paddy Graham, had announced the previous winter that he would be supervising a night school class in the Education Section, for the GCE in Air Navigation. Ten of us, mostly Officers and NCOs, signed up for it, and took the exam. Paddy must have been a good teacher for we all passed. The previous winter I had done O Level Mathematics, which gave me six O levels including a science subject. Alas, even with that, I still could not convince the RAF to let me through the navigator eyesight test.

Chapter 6 - Civilian Again

A few weeks after I left, I landed a job doing photo printing for a commercial photographer, but he wasn't doing very well so had to lay me off. Fortunately, just then, I was offered a job as a newspaper press photographer. That was a lot of fun, but the long-term prospects were not very attractive and, of course, I desperately wanted to get back to aviation. One of the jobs I had to cover was a weekend exercise with the RAF Stafford Mountain Rescue Team. That involved being lowered down a 115-foot cliff in the Pennines, strapped to a stretcher. My arms were free to use my camera and I got one shot with the guys at the bottom of the cliff and my old RAF marching boots all in focus. My editor loved that one— a huge pair of boots and all the tiny distant figures.

Our editor had a favourite saying, 'Faces sell paper', and always liked group photographs. I felt there was a rather boring sameness about so many of them, so injected a bit of variety into it. One evening, I was sent to photograph a group of schoolkids at a Christmas party and spotted a piano in the corner of the room. Nobody could play it, but I got one kid to pretend he was tinkling the ivories, and the rest of them to group round the jangle-box having a good old singsong. The editor loved that one too.

I went back to my old Air Cadet Squadron, Number 395 (Stafford) initially as a civilian instructor but was very soon commissioned into the RAFVR as a Pilot Officer. That was a lot of fun and good experience, in addition to being, occasionally, paid by the day for weekend attendances.

RAFVR School, White Waltham, 1964.

One time I was put in charge of the command rugby trails. I couldn't even tell you the names of the team positions, but that didn't matter. All I had to do was see that transport, food and accommodation for the cadets went smoothly.

In the summer of 1963 staying a couple of weeks with my ex-RAF friend Dave Bailey in Bromley, Kent, I took some more flying tuition with the Alouette Flying Club at Biggin Hill, but their two ancient Taylorcrafts were nearly always unserviceable and, before long, G-AFZI was written off on the Isle of Wight.[3] In fact I spent more time helping to fix aeroplanes that flying them, so soon lost interest again. In any case, there did not seem to be much sense in getting a PPL unless I could later get my investment back by progressing to a CPL and getting a job at a pilot—which the eyesight problem deemed impossible.

My mother had a bright idea. Most of the bright ideas in our family came from that direction. She was a lot more imaginative than my father who was a typically thorough, reliable, and hard-working NCO of the sort that are the backbone of the British Army. She discovered that I could work as an unqualified schoolteacher at half pay for two

---

3   G-AFZI, a 1941 Taylorcraft Plus D, suffered six separate accidents before being written off at Bembridge in 1963.

years, while gaining experience and waiting for a training college place.

I resigned from the newspaper probably just in time to avoid being fired. One black I put up was showing my two colleagues how to land from a parachute. All three of us had come hurtling down the stairs, taken a flying leap from about the eighth step, and rolled along the corridor just as the MD emerged from his office door with a pair of important clients. We didn't break any cameras doing that, we were just heading for a break in the coffee shop.

At the start of the 1964 summer term, I found myself teaching maths, English and a few other things I knew very little about to kids in the bottom two streams of a large seven stream secondary modern school. There was a severe shortage of certain teachers at the time, so some kids were getting more hours of tuition in singing than they were getting in English or Maths. A year of that was sheer torture. Kids never interested me, and I always made sure I never had any of my own. The little blighters played up dreadfully and I had serious trouble with class control. That was about the time that 'Six of the best' became unlawful, so the headmaster couldn't even use that deterrent. Certain other teachers were ardent trade union members and seemed to resent my unqualified presence, so they tended to adopt a very superior attitude and were not at all helpful.

One huge gentleman whom the kids nicknamed 'Basher' was quite unforgettable. There were too many staff for places on the stage during assembly, so some of us were stationed at intervals around the main hall. One time, just as assembly was over, he marched into the throng, grabbed two large lads who had been fooling around and banged their heads together hard enough for us to hear the thump. Later, in the staff room, as we complemented him upon this fine performance, he calmly observed 'Well, I couldn't really destroy them during divine service, could I?' Despite being a Labour councillor, I reckon he was one of the most ardent right-wingers I ever knew!

I had, of course, rejoined the Wolverhampton Aero Club

'With fire and heavenly thunderbolts thus may we speak with indolent and slothful senses. Thus spracht Zarathrustra.' Some of my favourite teenage thugs. Teaching school in 1964.

but was too hard up to do much flying at the time. Instead, I spent a lot of my visits tinkering around in the hangar and sometimes the manager slipped me a few pounds for helping out. If he wanted any photography done, he'd usually call me—I guess I was cheaper than hiring a local professional.

In 1970, the local council, as owners of the airport freehold, decided to close it down and sell the site for building land, forcing the aero club to close and all the privately owned machines to move to Halfpenny Green aerodrome. That boasted hard runways but not such a good weather record and was much further from the developing motorway system.

At school, dust from the chalk gave me a lot of sinus trouble and I was lumbered with classes of kids who should rightly have been taught by a specialised remedial teacher. But, at least the experience teaching there, and in the ATC, got me my next job, which was teaching air law, airframes and engines, and eventually the Link trainer, at a commercial flying school.

Chapter 7 - Flying School

The London School of Flying provided the flying facilities for the Elstree Flying Club and operated a government-approved commercial pilot course. I was an Assistant Ground Instructor, teaching basic airframes, engines and aviation law whilst also being taught how to operate the ANT-18 and D4 Link Trainers. There was also some opportunity to fly with students once their Private Pilot Licences were issued, allowing them to carry a passenger. So, if I were doing nothing else, I would often take the other seat in a Chipmunk or Auster. I became quite good at map reading so my presence was viewed by the company as slightly reducing the chance of my pilot becoming lost or violating regulated airspace. Sometimes the pilot got bored and handed control over to me. I was very good at flying the Link Trainers (due to many hours of free practice) so I had plenty of chance to improve my straight and level flying.

A little while later I had to supervise a group of Air Cadets on air experience flying at Cambridge, for by then I had arranged a posting to 23(F) Squadron at Bushey, Herts. Late in the afternoon, one of the AEF pilots, an RAF instructor from a nearby station asked if I would care for a quick flight. By then I was in the process of checking out as a navigator with British Eagle and told him so when he asked what experience I had. He offered to hand over control to me and I flew it straight and level for a few minutes upon which he commented that I ought to have been a pilot rather than a navigator. Very kind of him to say that, but he never saw what my landings might have been like!

The Board of Trade inspectors descended upon LSF for their routine annual visit and were unhappy that I was teaching all the aircraft technical subjects, and all the air law as well, despite having no formal qualifications in either. So, it was decreed that I would have to be replaced and LSF proposed to move me to the parent company, British Midland Airways, as an Operations Officer at East Midlands Airport. I was interviewed by the Chief Pilot but before anything came of that I applied for, and was offered, a job with British Eagle International Airlines as an assistant flight dispatcher at Heathrow. Since I knew that Eagle employed about forty Flight Navigators, that sounded a better proposition.

Sadly, although the school survived for decades, the Commercial Pilot course closed down a few years later. That was rather unfortunate because it had done a fine job and many of the graduates became Boeing 747 pilots, and at least one even made it to the Concorde as a captain. Unlike some other schools, cadets were never encouraged to 'play' at being airline pilots. The principal, David Ogilvy, always insisted we were training not airline pilots, but professional pilots. That way they would concentrate on thoroughly learning basic handling skills and good basic airmanship. The airline indoctrination for those going to the airlines could come later. The others would have the appropriate abilities for careers as instructors, agricultural, executive pilots or whatever else. Good thinking.

Chapter 8 - British Eagle

In fact, the only aviation qualifications I had were a restricted radio-telephony Licence, validated for ground use only, which I had acquired in order to help out in the Elstree control tower, and the GCE in Air Navigation, which I had taken on that part time course at Kinloss.

So, off I went to Eagle working in the LHR Queens Building on most aspects of flight planning and briefing crews under the supervision of Chief Dispatcher Alexander Torrance and his five other dispatchers. Soon I learned that the company also employed about ten cadet navigators who were all either ex-Merchant Navy officers, or RAF navigators without much long-haul flying experience (night fighter radar navigators for example). I also found out that the company needed a few more navigators.

In those days I was flatting with an old RAF chum (remember the Air Commodore's son?) who worked nearby as a technical civil servant. His immediate superior was a nice man but a rather reserved type with no sense of humour at all. He was always a bit 'distant' from the troops, but there was one girl he used to confide all his sorrows to. She, of course, went back to the crew room and relayed all the lurid details back to the lads.

The story came out that although their thirties were fast receding, they had, for some reason, been unable to start a family. His good lady—a hair in a bun, horn rim glasses, tweed suits and 'flat sensible shoes' type—and he made an appointment with a private fertility consultant. After

*Eagle Britannia G-ANCF awaiting us at RAF Changi, Christmas
1967. She is one of the three surviving Britannias.*

taking some notes, the consultant said he would require
a specimen. For this he gave them a test tube and said he
would lock them in his office for half an hour. They got
down to it and after about twenty minutes he was on the
finishing straight when someone outside rattled the office
doorknob. The result squirted right across the consultant's
desk so that when he came back, they were still frantically
mopping the last drops off the blotter into the test tube.

About that time my flatmate got involved another
amusing scam. He took some annual leave to work as a
film extra, in *Alfred the Great*, much of which was shot in
Southern Ireland. In those days all forms of contraception
were illegal there, so every time he travelled back, he took
a few huge 35mm film tins with him. Half of them were
crammed with French letters, which he sold at a huge profit.

By then I knew quite a lot about air navigation. Although
there was no advanced level GCE in the subject, the O level
was in, some parts of the syllabus, even more advanced
than the Board of Trade Flight Navigator exam. Trouble
was nobody in the airline business seemed to know that an
O level in navigation even existed. Now no longer issued,
the Flight Navigator licence was quite a tough exam,

being academically the senior civil aviation licence, and supposed to require about as much study as a BSc.

So having by then got to know Chief Flight Navigator, Reg Peake, and deputy CFN Howard 'Doc' Livingstone I suggested that I could do it. I was sharing a flat then with Mike Owen, a washed out RCAF navigator who had over 100 hours navigator time in his logbook (Mike had been obliged to leave over some monkey business involving the Commanding Officer's daughter). Later, he set up his own airline, which operated successfully for over ten years but eventually closed because they were unable to find suitable aircraft to re-equip with. He had the right to take the Flight Navigator written examinations thanks to that, but my problem was that the Board of Trade would not allow anyone to take the tests unless they had a minimum of 100 hours acceptable time as navigator under supervision. Mike was a rather better salesman than I, so he managed to persuade Reg and Doc to allow me to fly with training navigators during periods of annual leave until I had enough hours to do the test.

Reg did make one condition though. Before anything else I must go and get a Class One flight crew medical certificate from the Board of Trade. I was a little apprehensive about that because, at that time, all initial civil medicals were still done by the Royal Air Force. On the appointed date, I presented myself at No1 Central Medical Establishment, RAF, which was in those days was in Cleveland Street, West London, close by the Middlesex Hospital. When I arrived, the corporal receptionist on the front desk took my details and routinely asked if the RAF had previously examined me at any time. For once, without even thinking, I had the presence of mind to lie and say no. If I had been honest enough to admit that I had been three or four times, I am sure they would have delved into my records, found I had failed twice, and I would have failed again.

The medical took most of the day, and towards the end of the afternoon I was ushered into the presence of the Board President, an Air Vice Marshal. I was clad in absolutely nothing but a dressing gown, which I was told to remove.

Then he asked me to squat on the carpet as low as I could get. Next thing he told me to jump up into the air as high as I could, so I had a go and he seemed quite happy with the performance, adding something to his notes. After I had dressed, I was asked to screw my face up as tight as I possibly could. First attempt wasn't good enough, so he said try again. This time he was happy and scribbled some more notes. The reason for all this escapes me, but years later I discovered that the good Air Vice Marshall was a psychiatrist—so I guess that figures— they all go nuts in the end, or so I'm told! Many years later I learned that in wartime the AVM had saved the life of Catalina pilot, Flt/Lt Cruikshank VC, at RAF Sullom Voe, by improvising a blood transfusion before the grievously injured pilot could be removed from the wrecked aircraft.

He summed up by telling me that my right eye was borderline, but he was passing me since I only wanted to be a navigator. So much for the line that the RAF only made the measurements, and that the pass/fail decisions were the province of Board of Trade Doctors! Still, he took a reasonable view, and I remain very grateful to him.

A few days later, the official notification letter arrived, saying that I was acceptable as a navigator but not as a pilot or as an air traffic controller. That was fine by me, since that was all I wanted. However, they did explain that as age and experience, increase, the standard required for renewal is reduced. A few years later, when I decided to retrain as a commercial pilot, the medical people did not object. Consequently, I was able to fly as pilot with the rest of them for the final twenty-five years of my career.

On making my acquaintance with the Britannia it seemed navigation methods had changed very little since the end of the Second World War. The Royal Air Force Britannias had the navigation luxury of Doppler Radar giving drift and groundspeed readouts, but the equipment weighed a few hundred kilogrammes, which of course limited the available payload by the same amount. So, profit conscious civilian operators never used it, even removing it from the few ex-RAF Britannias that achieved civil certification. The basic navigation fit was two Marconi ADF radio compasses, two VOR receivers and twin ILS. Radar systems like Rebecca/Eureka and BABS had been planned but I never saw one fitted. Later, an early type of DME was retrofitted although I never saw one in a 102 series aircraft.

Within an airways network, or on initial climb, navigation was simple enough, merely tracking between VORs or non-directional (NDB) beacons, although, outside Europe and North America, there were very few of the latter. Australia had a very good system of Visual Aural (VAR) Ranges which had the advantage of requiring no other additional avionics fitted to the aircraft.

Operating in an 'off airways' environment the preferred basic fixing aids were Loran A or astronavigation (termed 'celestial' in America). Loran chains covered the Atlantic, Pacific, and parts of the Arctic and Caribbean but that was about all. An Indian Ocean chain had been commissioned in late wartime but was demolished in 1946 since the airlines didn't wish to finance it and the US Navy had no further use for it.

Over Africa and the Far East as far as Australia and New Zealand, Mercator charts were normally used. American 3071G Lambert Conformal charts were the norm on the Atlantic since they were overprinted with Loran and Consol information—Consol was developed from the wartime German 'Sonne' system. Charts at a much greater scale were also carried for map reading.

When flying within good Loran cover, fixing was usually at twenty-minute intervals, although in areas of poor cover, or lattice geometry, position lines could be crossed with Sun, Moon, star, radio or other position lines. At night an operator had to take great care not to cross a ground wave signal with a sky wave signal. It was a very useful system, although interpretation of the signals on the APN-9 set oscilloscope did take some degree of expertise. Unfortunately, most stations had been decommissioned by 1975 due to the increasing popularity of Inertial (INS) and Omega by the major airlines and military.

Britannias intended for UK companies still had a radio operator position on the flight deck, unlike those destined for the US market. It had originally been planned to fit a periscopic drift sight at the navigator's station. Those would poke out of the right side of the aircraft looking vertically downwards, but I never saw one fitted. At high altitude overland, they could have been quite useful but not over water or cloud. In practice, the nose radar could give a better indication of drift and ground speed.

Most Britannias had the roof mount for the Kelvin Hughes (Smiths) periscopic sextant, with which the navigator tracked the body for two minutes. The mechanical integrator took readings every two seconds so that at the end of two minutes the navigator could read the average of sixty shots which gave a pretty accurate result even in mild turbulence. In calm air, a practised navigator could usually get a three star fix accurate to within not more than seven miles and possibly less than five.

Inconveniently though, there wasn't a lot of stable air around on many routes, especially in the tropics where even the Boeing and Douglas jets couldn't get above all of the clouds. So quite a lot of shots had to be wasted and recomputed because the body disappeared as the aircraft entered cloud. That certainly kept one busy.

Five nautical miles (8 Km) doesn't seem very accurate by modern GPS standards, but it was a matter of 'make do' and, in any case, it was always good enough to home the aeroplane onto the destination radio beacon. If the beacon was out of service, one could always revert to Francis Chichester's 'Landfall' method by aiming slightly to the left or right so that if the airport didn't show up on time at least you didn't have to guess whether to turn left or right.

Using the RAF/US Navy AP327 sight reduction tables, an efficient navigator could compute, shoot, and plot a three-star fix in about fifteen minutes. Working in forty minute cycles, this left a comfortable twenty-five minutes for working out heading corrections, checking the fuel state against the 'Howgozit' graph, while keeping the air plot up to date and passing position information to the pilots on an AIREP form, sometimes adding weather information for them to pass on the meteorological office.

Daytime fixing on some overwater routes could be a problem. An early morning departure from Kuwait for Colombo meant that, at first, the Sun would be 'on the nose' but as all astro position lines are at right angles to the bearing of the body, its relative position moved to the beam as the hours passed. At the start of the flight, it was useful as a ground speed check, but later, conveniently, it could be used as a track check.

If the Moon, or sometimes Venus, was conveniently situated then a line could be crossed with a Sun line. However, if only the Sun was available and at forty-five degrees relative to the track, it could get difficult. Then one used a system based on simple statistics and combined the Sun line with the Dead Reckoning position to arrive at an 'MPP' (Most probable position.)

Powered by the curious reverse-flow Bristol Proteus engines, Britannias always flew at a constant power setting of about 11,500 compressor RPM in the cruise, and speed was controlled by selection of a suitable flight level. Depending on weight and outside air temperature, this usually allowed a 2,000 foot 'step climb' at about two-hour intervals. Before the 1973 fuel crisis, engine time was more expensive than fuel, so often we flew lower, just climbing as we reached normal operating speed limit (Vno).

After 1973. the long-range cruise was used almost exclusively. This meant we would climb as soon as weight and temperature permitted. It gave a lower Indicated Airspeed but a much better fuel mileage, even against a headwind.

Many of my former colleagues had originally been radio officers and we all held the same radio licences as the pilots. On long airways sectors it was quite normal for the navigator to use one of the front seats and give a pilot a couple of hours on one of the banks, if fitted. I could even work the engineers panel in the cruise and maintain the engineers log, which was useful experience when I later cross trained as pilot.

We didn't need much in the way of equipment. Just a pair of dividers which would double as a compass, keeping it all nice and simple. A straight edge and a Douglas protractor or a 'Weems Plotter' which was combination of the two. A supply of sharp pencils and a circular slide rule computer was just about all we needed although the early pocket calculators became quite indispensable. The 'drift side' of a navigation computer was rarely used even in the flight planning stage. The results were more accurate than we needed, and it took too long. We used a 'Drift Card', which was calculated for a fixed true airspeed in the climb, and another for a fixed airspeed for the cruise.

The first trip I did was a trooping flight for the Ministry of Defence to Nairobi and back. The aircraft was Britannia 312 G-AOVB, commanded by Captain Dixie Dean with a double crew including F/N Al Segal, who was supervising my efforts. We arrived at Benghazi late in the evening to be greeted by a voice from the darkness saying, 'Do you want fuel?' The skipper's reply was that he couldn't think of any better reason to arrive at 'This fucking dump in the middle of the night!'

We made it direct to Nairobi, in spite of my efforts with the sextant and the astronomical tables. It was only a minimum rest stop in the Nairobi hotel after which we set off for Khartoum. Nairobi is 5,284 feet above sea level, so even in the cool of the night we could not carry the payload, plus enough fuel to make Benghazi direct. That was in late December, so most of the flight was made in darkness, allowing me to collect a few astro sights towards the total I needed to get my licence issued.

Next trip was to Singapore and Sydney, with a slip crew commanded by the late Captain Pete Winslett. The first sector was a twelve-hour direct flight to Kuwait, where we took a rest whilst a slip crew took the aircraft on to Colombo. Twenty-four hours later we took over another Britannia from the inbound crew and carried on to Colombo. That was mostly a daytime sector, with only the Sun for position lines but, somehow, I found Ceylon parked in the right place, again with the aid of Al Segal.

Colombo to Singapore overnight was a tough one since the route more or less followed the Intertropical Convergence Zone so that we were in turbulence and in and out of cloud for most of the way. That didn't help my collection of astro sights, as half of those I had precomputed had to be abandoned when we flew into cloud and lost sight of the stars.

After a day in Singapore, we pressed on to Darwin, where we rested for another day, then on to Sydney and back all in one day. We retraced our steps back to Kuwait, but against the prevailing headwinds it was not possible to reach London in one hop. Westbound the service always called in at Istanbul for fuel and a crew slip. Since the navigator was not really needed on the Istanbul to London sector, navigators often remained on the aircraft so that they could get back early and have an extra day at home. However, I decided to stay with the crew and take a short look at Turkey.

We had about twenty-four hours there, so I took a train to downtown Istanbul with a couple of the girls for a shopping trip—they were not encouraged to go without a male crew member. We managed to take in a few of the sights, including a tour around the enormous mosque. I think the girls were June Phillips and Jackie Woodson. The ritual haggling with traders in the Middle East was always a good laugh. Usually, we'd get at least 50% knocked off the initial asking price if we bought any souvenirs.

I remember Jackie for another incident. We were having a few drinks in someone's hotel room, when she ventured to ask the captain, Australian John Nankervis, why that big wheel near to his right leg on the flight deck kept moving back and forth on its own. John explained that was the auto-trim, through which the automatic pilot compensated for people walking up and down the cabin. Then he jokingly observed, 'But when you walk up and down luv, it nearly takes me bloody leg off!' Jackie did have a rather ample figure! There was always a great atmosphere among us though, so she took the joke in good part.

John had a rather unfortunate experience. The hotel we

used in Istanbul had room doors which opened outwards into the corridor. Many of us used to sleep in the raw in the tropics and one night he woke to make a quick bathroom visit. Emerging therefrom he took a wrong turn and the door swung shut marooning him in the corridor. Although probably half asleep, he dealt with the emergency in an admirable manner. At the end of the corridor was the janitor's storeroom, so John found a toilet roll, wrapped it around his middle a few times to make an improvised kilt, then got the night porter to open his room.

A later trip took me to Singapore and Hong Kong via the same routing. Then, in 1967, things changed slightly when the India/Pakistan war cooled down and we were allowed to overfly India. We said farewell to Colombo and from then onwards routed via Bombay. That saved about one hour of flying.

In the nineteen-sixties, most airline crew were still veterans of World War Two and there seemed to be very few under the age of thirty in British Eagle. A handful had even flown as airline crew before the war. There was a scattering of gallantry medals among them, though few actually seemed to wear their ribbons. Most were modest men though, now and then, they would speak of Lancasters and Halifaxes, Mosquitos and Stirlings. One of them had even been a captive in Colditz Castle and another, a former Spitfire pilot, had been involved in The Great Escape. Later, I flew in the Boeing 707 with him. There were also a few Poles and Czechs whom, in later years, I discovered had impressive wartime records.

Being rather cautious, I didn't want to waste the massive examination fees (all of seventeen pounds!) until I was confident that I would pass. Mike Owen was a much more positive thinker and entered himself for the exams against my advice. When he got the grade slip, I was gob smacked: he had failed, but only by very few marks. We both immediately entered ourselves for the exams, and both passed in enough subjects to gain what was known as a partial pass.

I always felt that my weakest subject was meteorology, so

figured that I would concentrate on all the other subjects and hope I'd get a partial pass. Then take a month to really study hard at met and pass second time. Imagine my shock when I got my grade slip: I had passed on both met papers. and failed flight planning, even though it was my day job!

I really enjoyed much of the studying because I found it so absorbing, especially the astronomy and form of the earth aspects of the syllabus. In those days, the examinations took an entire week. All the question papers required essay style written answers with one paper each morning and another in the afternoon. They were held in the main hall of Burlington House, headquarters of the Royal Academy. There would be a dozen or so navigator candidates at the back, and the rest of the hall filled with pilots taking the tests for their Airline Transport Pilot Licences. We were also tested for flashing and aural Morse code at about six words per minute.

I soon retook flight planning and passed it. Then Eagle lost a big military trooping contract and had to lay off ten or so navigators and some flight dispatch staff, including me. Fortunately, Operations Manager Harry Wyatt—himself a navigator—and Reg Peake, agreed to give me the rest of my required flying, and a flight test, so that I would at least get a licence. Consequently, I did a final couple of trips, which bought my total time above the minimum I needed for my licence. On the last trip I had already left the company and was on the payroll of Britannia Airways, but they kindly gave me the time off to complete the project. My last flight with Eagle was in Britannia freighter G-ANCF from RAF Changi to London. George Rose was the first navigator, but apart from appearing on the flight deck a few times to make a fuel check, George left me to myself. Singapore to Bombay and onward to Kuwait were the first two sectors I did on my own.

I also had to take the 'Performance A' test: a problem, because there were no proper training courses for it in those days. It seemed the only way was to find someone who had done it himself and get him to show you how. Fortunately, my old boss at Elstree, chief ground instructor

Sidney Taylor, kindly lent me some BOAC training notes, which got me through the test at about my third attempt. At the back of my mind was the hope that business would improve, and I would be able to go back to British Eagle, but it was not to be. Tragically, in late 1968, the company closed and nearly three thousand loyal employees lost their jobs. The Labour government simply did not want any competition against the state-owned BOAC and BEA. Eagle had three main profit centres: inclusive tour Mediterranean holiday flights—a market the company had pioneered; military trooping contracts, and scheduled services. Founder and chairman, Harold Bamberg, hoped that if one of those failed, the other two would keep the firm going. Unfortunately, the trooping contracts waned as British forces began to pull out from 'East of Suez'. Then the Wilson government devalued the pound and imposed a fifty-pound limit on foreign holiday expenditure. That had a detrimental effect on the inclusive tour market, from which, some airline economists have argued, it has never entirely recovered. With two out of three profit centres gone, the scheduled services—though just beginning to show a profit—were simply not enough to keep the firm afloat.

Inefficient operation and bad choice of aircraft types had left both state corporations in a woeful financial condition, and around that time they had to be rescued by an input of about £110 million pounds from the taxpayer. By contrast, Eagle was allowed to go broke, a mere two million or so in the red.

Eagle had flown for twenty years, and along the way made an enormous contribution to the Berlin Airlift—for which Harold Bamberg was awarded the CBE—and to UK civil aviation. The company had a wonderful spirit, never quite rivalled in any other firm I've worked for. Fifty years on, we are still having annual reunions.

Chapter 9 - Britannia Airways

One door closes and another one opens. On the very day I found out I was redundant, we happened to be handling a Britannia belonging to Britannia Airways, of Luton Airport. It so happened that the first officer was an Australian, Brian Johns, whom I had known as a flying instructor at Elstree. Learning that I had just been put on three-month's notice, he suggested that I get in touch with Britannia, since the only navigator they had—other than the chief—was actually a captain. He was due to go onto their new Boeings and would not be available to navigate Britannias for much longer. Interviewed by Chief Pilot Derek Davidson and Chief Navigator George Beresford, I was eventually offered a job, doubling as personal assistant to Commercial Director Bob Horlock at his London office, when not required to fly trips.

Of course, I only had just over two hundred hours of 'N2' logged as navigator, and almost no 'N1' as first navigator, so I was not exactly navigational hot property, but this was a good way to build up some experience. Most of the flying was inclusive tour holiday work to the Mediterranean, which did not involve me, but we picked up some long-haul charter work around Africa. Some of the first flights I did were to Lagos, Nigeria during the Biafra War. The Nigerian Federal Government had the bright idea of replacing the currency at short notice, so that all the Nigerian notes the Biafrans held would become worthless. They had thirty-seven tonnes of notes printed in the UK and we were contracted to make five flights to deliver them. Our aircraft were the short-range Britannia 102 series, with only four fuel tanks, but by stripping out all non-essential equipment, we could operate Palma to Lagos direct with just over ten tonnes.

When the first flight arrived in Lagos we rolled to a halt on the airport ramp and the air stairs were pushed up to the forward door. We were then confronted by a soldier with a cocked Sten gun, something I was familiar with from Cyprus EOKA days. He asked if we had any Nigerian money on board! Silly question, I guess, but before I could respond with, probably, an equally silly answer, the captain, Andre Jezsiorski (who was said to be a Polish count) appeared behind me and peremptorily ordered the soldier to get off the aeroplane and come back on properly. That is, with his gun to safety and not pointed at us!

At the time we arrived, everyone knew that the currency was about to be changed. Anyone who had previously taken Nigerian money out of the country was frantically trying to smuggle it back in and Nigerian Customs were equally frantically trying to stop them!

Unfortunately, we had no back loads arranged, so were able to ferry back to Luton nonstop, which involved quite a lot of revision of the initial flight plan in the air. The technique was to keep the flight legal by initially planning and filing a flight plan for Palma. Then, if we had gained some fuel and time after about four hours, I would re-plan and see if we could make Paris/Le Bourget. After another two or so hours I would re-plan again, this time for Luton. I think we only had to land at Le Bourget once. That was long before the days of pocket calculators, so I was working away 'number crunching' with my slide rule for hours on end.

In those days, most airline stewardesses seemed to smoke, so I was more than delighted when a lovely young honey blonde who didn't smoke—never mind her name—began to take a rather more than casual interest in me. Things were going incredibly well when I found she was flying with me on a trip to East Africa. Layovers were usually quite short, but for some reason we had a couple of days in the Lake Victoria hotel at Entebbe, Uganda. The afternoon was hot and humid, so were lounging around half naked on my bed. Both lying on our backs and kissing voraciously as she expertly manipulated my wedding tackle. Alas, I lost control of the situation, and the resultant emission scored a direct hit in the middle of my right ear. I doubt if that has often been surpassed in the annals of sexual science.

Not long after it was decided it was time for me meet the parents and I was summoned to attend the family estate for a weekend. My car was a bit of a wreck, good enough only for leaving for days in an airport car park, so it was decided we would travel in her Mini. By then, I knew that her father was a wealthy solicitor, but as we swung into the driveway of the parental mansion—which probably had more than ten bedrooms—I began to experience a serious lapse of confidence. Her mother seemed okay, a quite homely, friendly and lovable 'mum', but 'Daddy' was quite something else, a rather diffident and humourless individual. I sensed from the start that he didn't really think much of me. At one stage on the Sunday morning, I was invited into 'the presence' and after a few brief enquiries he got around to the Victorian question of, 'What are your intentions regarding my daughter?'. Fortunately, I managed to stop myself coming up with some such fatuous retort such as 'Thoroughly dishonourable', but I just managed some weak-kneed reply such as 'She is a lovely girl, but really it is up to her, I suppose'. I can't recall the wording of his reply, but it went along the lines that the career prospects of airline crews, especially navigators, were unlikely to provide his daughter with the lifestyle he intended for her.

I really put my foot in it at dinner that night. There were some visiting relatives which meant there were about ten of us at the table, and the conversation got around to politics. The lavish surroundings had inferred conservative sympathies, so I made some derogatory remark about the mess the Wilson government was making of things. Instantly, the conversation halted, replaced by a crashing silence. After an interval of a few seconds, her mother made a faltering attempt to restart the conversation. Eventually, the dinner party broke up and the pair of us took a stroll down the garden. There she explained that 'Daddy' had been an unsuccessful Labour candidate in a recent by-election and was on first name terms with Mr Wilson. Oh dear!

I had to get back to work on the Monday, so she dropped me off at the local station. Some paternal pressures must have been applied because she was very quiet and that was the last time I ever saw her. She was leaving the company to take a job with BOAC anyway, but next time I went round to her flat, the other girls told me she was away. After a few more unproductive visits and phone calls, I wrote to her, but she never replied. Maybe a few bunches of flowers would have done the trick, but I wasn't smart enough to know that in those days!

My work in the London office was varied and interesting, giving me a very valuable insight into the commercial aspects of air transport. Often, I would take an initial charter enquiry, analyse it and calculate the flight times and payloads, consult with accounts over pricing, work out the schedule and make inputs to the 'Captain's Brief', and then actually operate as navigator on the trip. Unfortunately, I had to waste a lot of time dealing with time wasters and young hopefuls trying to make some money out of charter flights, for that was in the years before licensing of tour operators and before the ABTA bonding scheme.

Later, we picked up a passenger charter, this time in the other direction. It involved an empty ferry direct to Lagos to collect a party of German tourists from a cruise ship. A British Eagle Britannia and a Caledonian Airways Boeing 707 had already been positioned out there and we carried both their slip crews out as passengers. At that time, I was dating Barbara, a German girl on the Eagle crew. The number two stewardess, however, was Anne Harrison, whom I knew slightly from the past. Oddly enough, over the years I had worked for Eagle, I had seen Anne checking in for flights a few times but had not been entirely sure it was the same girl. After all, air stewardess training, along with company uniform, hairstyle and make up rules, do change the look of a girl. Likewise, she did not remember me. Then we happened to be sitting on adjacent deck chairs round the hotel swimming pool, when I overheard here mention the word 'Stourbridge' to another girl. The penny dropped and each of us realised who other was. After we left Lagos on two different aircraft, we went our separate ways and never met again until Anne discovered I was based at an airport close to her home and contacted me in 1997. By then, we had both been married and divorced. We have been together ever since.

We were notified at, one stage, that the cruise liner was late and that there would be a twelve-hour delay. All the crews repaired to the swimming pool although, mercifully, the bar was closed. Suddenly, my own skipper, Captain Alec Crawford, appeared at the poolside resplendent in his uniform, waving a telex signal which said we were to leave immediately, as the ship was in port. The girls all assumed this was a wind up and were very close to throwing Captain Alec into the pool—uniform and all. Fortunately, sense prevailed and there was a headlong rush back to our rooms to change.

Not long after that I became involved in some monkey business over Rhodesia's Unilateral Declaration of Independence (UDI). Air services between the UK and Rhodesia were suspended, making it impossible to arrange passenger charters between the two countries. A charter broker contacted me about the possibility of some flights between Beira, Mozambique, and Dublin. Allegedly, these were for expatriate Irish living in East and South Africa. We fixed up two flights, which I duly operated as navigator. The first was an empty southbound leg via Benghazi, Libya and Entebbe, Uganda. After a rest period at a hotel, we returned to the airport late at night. No passengers were to be seen, but soon the whine of Rolls Royce Dart engines split the night, and two Vickers Viscounts in Air Rhodesia markings emerged from the darkness. All of my paperwork was complete, and I was ready to go but first the passengers had to be processed and tickets issued, etc. While this was being done, I sat on the edge of a baggage cart in the cool African night chatting with one of the Air Rhodesia cabin crew. She was a charming blonde named Karen, who had only begun flying a few weeks before. A few months previously she had lived a few miles from me in Pont Street, Kensington. If only I'd known! She was one of the most delightful girls I have ever met but I never saw her again. I just hope that she survived all the subsequent trouble in that sad country and that life has been kind to her.

Northbound, we stopped for fuel again in Entebbe and Benghazi, and duly delivered the passengers to Dublin. Two days later, Luton Operations phoned me to say that the shit had hit the fan and there was talk of an 'unlawful' flight. Questions were also due to be 'asked in the house' about it. To my immense relief the hue and cry died down and was soon forgotten. Whew!

There was some more trouble around that time, which thankfully didn't involve me. A local newspaper ran an article about dodgy charter airlines operating from the airport. What the editor sadly forgot was the fact that the group that owned his paper also owned us! He was soon called to order by the top brass, so was the journalist.

On one occasion, I left a northbound service at Benghazi to take another one south a day later. In the bar, I met up with John Field, a former RAF Pathfinder Force navigator who was now Chief Navigator of African Safari Airways. They operated a single Ugandan registered Britannia 300 on passenger work between Europe and Kenya. He offered me some freelance work, which I was able to accept during leave spells. All the work was between Gatwick or Basle and Nairobi, except for one flight to Antananarivo (Ivato) airport, in Madagascar. The trip length involved a double crew with both John and me operating as navigators. On arrival, we were placed under house arrest because there was some question over traffic rights. Happily, we were 'banged up' in a fine hotel although the French food didn't appeal to me: bouillabaisse soup with fish heads floating around in it! When the problems were sorted out, we were able to leave with our northbound load, but it had been hard on our two Kenyan cabin staff. They had not been allowed ashore at all and were forced to sleep in the aeroplane.

Another unusual request I had to deal with involved the famous Belgian mercenary fighter 'Black Jack' Schramme and his troops. They were holed up in Bukavu, surrounded by hostile forces and desperately needing to be rescued. I was asked if we could send an aeroplane to collect them, but it was a non-starter in every way. Our shorter range Britannias could not carry enough fuel for a round trip and to carry our own fuel in drums in the cabin and refuel from them would have been too risky and taken far too long. Most other African countries would not even allow us to overfly, and in any case, I think our group management would have vetoed the project. That would have been fortunate for me, perhaps, as I knew quite well who would have pulled that trip as navigator. Some other company did make a successful rescue, but certainly with a more suitable aircraft and probably breaking a few laws in the process.

Britannia was always a respectable airline, but from time to time we did engage in arms and ammunition delivery flights. I suppose some would have categorised us as 'gun runners' but all of these were completely above the board and done with full British government approval. The cargoes came from Royal Ordnance Factories at Enfield and our major customer was the Federal Nigeria government. Most of we crew members had more sympathy with the Biafran cause, but sadly felt that they were no-hopers and had no chance to win. Reluctantly we felt that the UK Wilson Government was right in taking the view that if we continued to supply arms it would shorten the war and fewer would starve and die in the end.

Naturally, the French took a different view, since they were supplying Biafra and hoped that if the Biafrans won, they would be able to get valuable concessions in the oilfields of East Nigeria. We were denied overflight clearance by France and had to route out over the Atlantic and the Bay of Biscay before turning inland over Spain at Bilbao and landing for a fuel stop at Tripoli. This incurred a very long duty for me so, on the first leg, I used to lay an inflatable mattress on the ammo boxes as we taxied out so that as the cabin pressure fell, it would inflate itself. I would then grab a few hours in a sleeping bag before flattening the mattress at top of descent, so that it would roll up nice and tight when air pressure flattened it.

The northbound flights were empty although sometimes we did bring a back load of fruit and the French did allow overflight for those. For the most part, we made Manston direct. Late in 1968 two other firms, Lloyd International and Air Spain, were tasked along with us in moving two hundred tonnes of ball ammunition from Manston, Kent, to Lagos. At the other end of the ramp another airline was taking ammo—I assume from the same source—to Fernando Po for onward shipment to Biafra. Figure that out for yourself.

The UK government loaned a huge ceremonial sword to the Nigerians for some parade, maybe associated with their independence jollifications. Just as we were about to leave, an Embassy official came out to the aircraft and asked if we would take it back to London. This was a two-handed affair, a real old Excalibur about four feet long. We had overflight clearance for Algeria, but on condition that they had the right to call us down for inspection. Fortunately, they never did, since I was a bit worried we might get arrested for carrying 'munitions of war' on what was billed as an empty flight. As I worked in the London Office between trips, the skipper suggested I should take it with me when we reached Luton. No problems with the UK Customs, so it came home with me (in a box), and later to the office, via the Piccadilly Line. It sat upon my desk for about two weeks until I had a call from some Rupert in the Foreign Office, who eventually sent a cab round to collect it. For a couple of weeks my desk boasted what was surely the most impressive letter opener in London.

A number of flights from Uganda to Luton were carrying Asians, who began to leave East Africa long before Idi Amin overthrew the Milton Obote government. In those days the Israelis were training the Uganda Air force. Security being very tight, I never dared be seen with a camera on the Entebbe ramp, which was a pity since we were once parked right next to a lovely old de Havilland Dragon Rapide. There was some uproar in the newspapers back in the UK when it was discovered that our passengers had been met at the airport by taxi touts who offered them rides to London and dumped them on a street in downtown Luton, telling them they were in London and charging them accordingly. Those cab drivers—sadly— were also Asians.

One aspect of navigating a Britannia was a risky business. Using the persicopic sextant you had to stand on a stool which retracted mechanically into the floor when not in use. With two hands working the sextant, one was very vulnerable to certain stewardesses who would creep up behind and start removing your trousers. Still, I guess one cannot complain too much about being 'debagged' by such beautiful young ladies!

In July 1968 we took delivery of our first Boeing 737 aircraft, ferried non-stop from Goose Bay, Newfoundland and flown by our chief pilot Captain Derek Davidson. These were short haul aircraft only, which meant that once the Britannias were gone, my flying career as navigator would come to an end. However, I did get involved in the 737 operation in some ways. As part of the certification process, the firm had to demonstrate emergency evacuation to the satisfaction of the Board of Trade. In those days, unlike today, it was okay to use our own hangar and office staff. I became one of the first few dozen British to evacuate from a Boeing 737 down the emergency slides. Sadly, as the aircraft had little fuel on board, the slides were steeper than usual. I got away with it, but three unfortunates were taken to the Luton & Dunstable Hospital with bruised vertebrae.

One of the senior captains was rather an oddball. Though formerly a highly experienced and qualified engineer, he resented the presence of anyone who knew more about something than he did, so especially didn't get along with the Chief Navigator. At times he was cantankerous enough to be quite a risk. If he was advised to turn a few degrees one way, he would intensely study the pitch dark through the windows for a minute or so, and then turn the other way. One night they just levelled off leaving Benghasi, Libya. He was fiddling with the radio compass and the needle swung round and settled pointing dead ahead. He observed 'Who needs bloody navigators— that's pointing at Karima already' (about a hundred miles north of Khartoum and at least a thousand miles away). By the time the navigator could reply 'I beg your pardon', the needle had swung round towards the right wing tip and stayed there for the next four hours or so. My boss made sure I was never allowed to fly with him, so the old boy never knew he would be getting no more trips outside Europe. People can be strange and, oddly enough, he seemed to mellow with age for when I met him at an air show a few years before he died, he seemed to have become a nice old chap and we had quite an enjoyable chat about old aeroplanes.

An early revenue B737 service needed an engine change in Barcelona. It fell to me to find a cargo aircraft to take the new engine out and bring the old one back along with the transit stand. Nobody had any availability though I even went as far as trying Bristol Freighter owners. In desperation I rang the USAF at RAF Lakenheath, who were most helpful. They put me onto a USAF Colonel in Frankfurt who said they were happy to do it as a training exercise and would charge the IATA commodity rate for light machinery, subject to written legal assurances that there was no civilian operator here or in Spain who could do the job. This was agreed, whereupon the unofficial contact had to be replaced by a formal application via the Foreign Office. By the time we were half way through those 'usual channels', word reached me that it had been an electrical snag after all and the engine didn't need a change. We thanked the USAF, but it was nice to know they were there, should we ever need them in the future. Neat piece of USAF support for an American product, I guess.

Ship crew change charters were a feature of those days and were most welcome for they often took us to unusual destinations. In December 1968, we picked up a crew in Durban, South Africa. It had been requested that we provide plenty of beer on the service, which we did. There were only about fifty passengers and the fact that they must have come from a 'dry' ship, along with the pre-Christmas spirit, ensured that the bar had been drunk dry on the first sector, Durban to Entebbe. But there was a good atmosphere in the cabin, nobody made a nuisance of themselves and for the sectors on to Benghazi and Luton most of them slept it off, so we didn't have to re-stock the bar.

I picked up four flights to Colombo, Sri Lanka but only operated on the final one. The first leg of the return journey was Colombo to Dubai and that was a critical sector for a short-range Brit. On one of the earlier flights the pilots had some difficulty in trimming the aeroplane in pitch but could not return to the airport, as it was way above maximum landing weight. The problem was not deemed serious enough to dump fuel but at top of climb they needed almost maximum nose down trim to remain straight and level. This slowed down the cruise speed slightly and the flight only just made it to Dubai so, at that point, it was decided to investigate what was in the rear baggage hold. The passenger bags had not been individually weighed but an average weight assumed, as is usual practice. The offending item was eventually discovered. Someone had loaded a cabin trunk containing a complete six-cylinder auto-mobile engine. So far back from the aircraft datum, it had caused the centre of gravity to move dangerously outside of limits.

When I was rostered to do the final trip of the series, the plan was to fly a 'slip crew' out to Dubai, via Beirut, Lebanon. We were due to leave Heathrow on a Middle East Airlines service, then hop a Kuwait Airways Comet via Abu Dhabi to Dubai. I was about to leave my flat, when operations called to say wait until they called again.

Some hours later we did leave and arrived in Beirut that night. And what a mess Beirut Airport was, the ramp was scattered with burnt out wrecks of airliners, for the Israelis had been raiding the airport just as we were due to leave Heathrow. Actually, I was rather sorry I'd missed the show, by exercising 'bragging rights' I guess I could have dined out on that for years. At least I don't think any passengers were harmed in the raid.

At Colombo they had recently built a brand-new terminal building and we were informed that there were ten comfortable rooms for transit aircrews on one of the upper floors. Tired, I opted to take one of the rooms, not fancying the long ride in a decrepit bus to downtown Colombo, about twenty miles away. The captain and the girls all headed for town leaving the co-pilot, engineer, and myself. We had a few drinks in the engineer's room and then I headed to my room. Alas, I had left the door open and a light on. It had got dark so that the room was infested with hundreds of local bugs. It certainly would have delighted any entomologist, but I had to get a move to another room or get eaten alive.

A frequent customer for cargo flights was Canford Aviation Services. Their managing director, Alan J Stocks, was brilliant at getting business and had many contacts around Africa. We often took a load of Phillips electrical goods from Rotterdam to Lusaka or Ndola, Zambia. On arrival it would not be unusual to see the African loaders dragging huge cardboard boxes clearly marked 'Handle with care' and 'This way up', upside down, down the air stairs – bump-bump-bump. By contrast, loaders in Asmara, Ethiopia, where we often picked up a back load of green peppers, were brilliant. They formed a human chain, passing boxes along, and once they began singing their work chant, nothing could stop them. Aircraft were loaded in record time, and we always enjoyed our stay there. The motel was very modern and clean, the food good and the water fit for drinking, unusual in Africa. The rarefied air at that altitude took a little getting used to. Happily, the motel was all on one level and there were no stairs to climb. Canfords eventually acquired their own Britannia and were renamed International Aviation Services (IAS), growing over the years and finally taking over Trans Meridian to become British Cargo Airlines which sadly closed down in 1980. By then I had been able to do a number of freelance flights as navigator for them, although their later DC8 aircraft were INS equipped. Years later watching some of the clowns here in the UK and Ireland loading or unloading cargo, I often though of those Ethiopians. They'd have made a much better job of it.

Back in 1964, a Dutch de Havilland Dove visited Wolverhampton on a charter flight. I photographed it but soon forgot it. Then in, mid-1968 a feature appeared in the *Air Pictorial* magazine, all about that Dove operator, Martins Air Charter. In a mere six or so years of operation it had grown to be a substantial company which include four Douglas DC7C airliners operating long range charters all over the World. This looked like another good opportunity for some freelance work, so I sent them a short telex message. A few days later I received a letter at home saying they needed a freelance engineer for some Atlantic DC-3 Dakota flights. Clearly this was an error, since the Atlantic would have been a little too much for a fully loaded Dakota. I queried this and they came back saying it had been a mistake, and please could I do four DC-7C trips to Paramaribo, Surinam. Immediately a telex signal went back, saying 'Yessir! Yes Please!' But then I guess some unfortunate Dutchman had a quick fag in the wrong place and a hangar at Schiphol was burnt down destroying two aircraft. This meant that KLM were not able to release a second DC8-33 to Martins which disrupted their entire programme to the extent that the Surinam flights had to be sub chartered. In fact, they were sub chartered to British Eagle and Anne flew on the very last one, which tragically was Eagle's final flight into Heathrow. I lost that work but, fortunately, Martinair kept my details on file.

In January 1969, Aer Lingus interviewed me for a job on their Boeing 707 Atlantic routes. Britannia wanted me to stay on permanently, as a commercial executive, but with over a thousand hours as first navigator, I naturally wished to remain flying for as long as I was able. I had overcome a lot of obstacles to get my licence so it made sense to keep it, if I could. In fact, so far as I am aware, very few other 'self-improvers' ever got a British Flight Navigator Licence without previous experience as a foreign or military qualified navigator, a Merchant Navy deck officer, or a retrained airline pilot or radio officer.

Martinair called me again and asked me to attend an interview at Schiphol, kindly sending me a return ticket, so off I went. The interview, with Chief Pilot Henk Fransen and Chief Navigator Jan Jager, went well and I was offered a seven-month contract on the DC-8. The job market was fairly positive at the time, and jet experience was a big plus. Figuring that I would not have much trouble getting another flying job in the autumn, I accepted on the spot.

The day after I returned, I received a delayed job offer from Aer Lingus that had been sent to the wrong address and then forwarded to me. I had to reject that and give my notice to Britannia. They did not take it very well, and I received unpleasant phone calls from both the chief navigator and the deputy chief pilot. I gave three months' notice and was due to take the final two weeks of my unused annual leave at the end of it. I had quoted for a flight that would have needed me, but I knew the guy involved was a daydreamer and that the flight would almost certainly never operate. On the other hand, I needed to go to Amsterdam to do some conversion training during those two weeks. Unfortunately, the deputy chief pilot got wind of it and tried to cancel my leave. I insisted on my rights and my immediate boss, the commercial director, backed me up. The flight never did operate because the destination country refused to grant traffic rights. But Britannia was a good airline where things were done properly without cutting corners. All in all, I was quite sorry to leave, even though it didn't have quite the atmosphere Eagle had.

In fact, my boss had been rather annoyed at the deputy chief pilot for interfering. We made a point about keeping very much to ourselves the details of what we had quoted for. It was much better for everyone if Flight Operations Division knew nothing until the contract was signed, otherwise they were likely to waste a lot of time and effort planning for flights that would never happen. We also kept all financial costings and charter prices a closely guarded secret for we did not want crew members doing their own arithmetic and trying to assess how well the firm was doing. That could only cause trouble since there was no way they could ever hope to get their sums right. Too many factors were involved which would be unknown to them.

A few years later Britannia did re-enter the long-haul charter market with a pair of fan-engined Boeing 707s, but the project didn't work out and they were soon leased out before being sold. Many former British Eagle pilots joined the company, which continued to flourish eventually becoming the world's biggest charter airline. It continues to this day, now trading under the name TUI Airways.

Chapter 10 - Martinair Holland

Captain Klaus Dykstra commanded the DC8-33, PH-DCD, when I did my check flight to New York, with Chief Navigator Jan Jager. Both real gentlemen with a wonderful dry Dutch sense of humour. The DC-8 was fitted with Doppler radar, which could sense drift and groundspeed and also had an Edo345 LORAN A receiver. Track keeping over the North Atlantic was very important and most of the work involved LORAN fixes at about twenty-minute intervals in order to update the cross-track error on the Doppler computer. The technique was quite different to the air plot methods used in the Britannia, and astro navigation was not used very much, except for a few sights in mid ocean on the overnight return trip. Of course, at Mach .82, the DC-8 was a third faster than the Britannia, but I soon got used to the higher speed.

Mr Jager was a firm believer in keeping everything as simple as possible (as Reg Peake had been in Eagle). He developed a simple and straightforward system of free gyro steering for polar routes. After one round trip to Vancouver BC, he signed me off, and at last I was off on my own with a DC-8. In a way, I guess, I was pushing my luck. One day I was heading west over the middle of Greenland, when my skipper looked round at me and observed that if I died right then he would have no idea whatsoever what direction he was flying in!

I would usually arrive at Schiphol airport crew centre two hours before scheduled departure time to collect the 'Met' folder and 'NOTAMS', then get an accurate time check. After a quick breakfast snack of coffee and Dutch apple pie at a 'Coffee Corner' in the passenger terminal, I would make a start on the flight plan. Computerised flight plans were already available in those days, but we felt that our manually produced ones were good enough. Routings varied at twelve-hour intervals on the North Atlantic Organised Track System (OTS) to get the best out of the changing upper winds. Weather data was usually pretty accurate so, after about twenty minutes with my 'whizzwheel' circular slide rule, I'd have enough information to fill out and file the ATC flight plan. A few more minutes and I would figure out the trip fuel and pass that data to the captain, who would complete the load and trim sheet.

Block to block, New York normally took about eight hours, entering British airspace at the Ottringham beacon near Hull and leaving somewhere like Skipness beacon in the Western isles of Scotland. Another of my duties was to call Shanwick Oceanic ATC centre on the VHF radio and negotiate our clearance into Oceanic airspace as we passed over northern England. The variable track was usually south of Iceland and the southern tip of Greenland entering Canadian air space maybe near Goose Bay, Labrador, then following Canadian high-level airways and American Jet Routes on to John F Kennedy airport.

On westbound daytime flights, astro navigation was rarely used for other than an initial heading check using the Sun. Primary navigation was by Doppler radar, which sensed ground speed and drift, feeding the data into two analogue computers. My main function was to take LORAN fixes at about twenty-minute intervals and use them to update the computers. Sometimes we could get a radar fix from a weather ship, one of which was sometimes manned by Dutch seamen. The skipper would allow the sailors to chat up our stewardesses over the VHF for a few minutes, although I'm not aware of any marriages resulting from that.

A Martinair DC-8 crew. NDW is on the far left.

Steering on the Atlantic was by gyromagnetic compass, but up north in the region of compass unreliability we used 'Free Gyro' steering. Close to the poles the horizontal magnetic force becomes so weak that the magnetic compass is useless, and the system is selected to run as a free gyro. In this mode it is affected by four types of drift, sometimes called 'wander' or 'topple'. The first three types of gyro drift (not forgetting wind drift of course) are caused by the Earth's rotation, movement of the aircraft over the chart and its movement over the Earth. These three are mathematically predictable using graphs or tables and allowed for by adjusting the latitude on the compass controller. Unfortunately, the remaining small error caused by in-service wear or manufacturing imperfections in the bearings causes a drift that cannot be predicted. For that reason the heading is checked by taking a bearing on the Sun, or a star at night, at intervals of not more than half an hour. The rate of drift is then assessed and that gives some idea what it will be for the next half hour. In periods of twilight, with no celestial body available, we used a Kollsman Sky Compass, which used polarised light above us to give a remarkably accurate heading check.

Polar navigation certainly has no room for errors.

Diversion airports are few and far between, navigational fixing aids are rare and, in an emergency, survival prospects are adversely affected by the severe cold. Indeed, many airlines included two navigators in each crew in the earlier years. It was important to keep navigation procedures as simple as possible to reduce the chances for error. In a region with little else as a directional reference, failure of a gyro compass system would be a serious emergency, and although the probability of it happening was very low, a procedure to cover double gyro failure was developed.

Our standard navigation plotting chart was the Polar Stereographic Projection, printed by KLM airlines. On this type of chart, a Great Circle is not quite a straight line but is close enough in practice. It has the great advantage that a difference of longitude on the chart is the same as on the Earth, and that reduces the chance of making a serious error in celestial heading checks. The chart was overprinted with a grid of lines parallel to the Greenwich Meridian, and all navigation plotting was done with reference to those, and not the true lines of longitude.

We felt it was good PR to allow passengers onto the flight deck in those pre-terrorism days. Some were interesting to talk to, especially those who had been aircrew in World War two. A few, especially private pilots, tended to outstay their welcome and insist on telling us all about their flying experiences. We devised a system of signalling the stewardesses to find some excuse to come and remove them discretely. After a while, I persuaded all the captains to deny access when we were steering by free gyro. That was because I was standing up using the sextant more than usual, and there wasn't room for standing passengers when I was doing that.

Martinair was the first non-scheduled charter airline to over-fly the North Pole, but our flights rarely exceeded latitude 85 degrees north. Fixing up there was quite difficult, signals from LORAN chains further south were very poor, especially over land in daytime. Nor were there many radio beacons, although we could sometimes get a radar fix from one of the DEW line radar sites.

SFO, later Captain, Peter Brown, in the right-hand seat of a Zambia Airways DC-8

For topographical charts we carried a set of US Navy 'Victor 30' maps. These somehow managed to give an incredibly good representation of the frightening complexity of the thousands of small lakes covering so much of the Canadian north. You could soon recognise the general outlines of the larger lakes with curious names like Yathkyed, Dubawnt and Ennedai.

The most worrying times were when flying between cloud layers, usually eastbound from Vancouver and still too heavy to climb above them. But those Sperry gyros were so good that I once flew more than three hours without a fix or a heading check yet found we were only seven miles (11Km) off track at the first available fix. Gladly, it didn't often happen, for cloud above 30,000 feet is not common at those latitudes.

Inbound from Europe we normally rejoined the Canadian airway system at Fort Smith, Fort McMurray or Uranium City, still steering about 340 degrees on the Greenwich Grid. At my final heading check, we realigned the gyros to the magnetic heading, somewhat unnerving for the pilots who would now find themselves steering about 190 degrees without altering heading.

The DC8 cruised at Mach .82, that is to say, 82% of the

speed of sound, which worked out as a true airspeed of about 480 Knots, varying slightly with outside air temperature. Even in conjunction with our somewhat unsophisticated flight planning methods, it was not unusual to hit the final approach beacon at Vancouver or Los Angeles right on time and with a fuel burn within 100 Kg of our estimate.

The big, fan-engined DC8 aircraft certainly had 'long legs'. Direct flights over about 4,850 nautical miles from Amsterdam to Los Angeles with a full load of 182 passengers, were routine. We didn't arrive with a lot of fuel, but that was not a problem; for the final three or so hours of cruise we were well within range of half a dozen airports big enough to take us. Arrival was always scheduled for early afternoon by which time any morning fog had cleared. Return to Amsterdam was more of a problem, even arriving late morning. European weather not being quite as reliable, we normally took a more southern route to pick up the tailwinds, and then tanked up with plenty of fuel at Bangor, Maine. At one time, Martinair had quoted for a flight over the South Pole, the navigation office producing our own charts, but the flight did not operate. It would have been too expensive, and the charterer could not sell enough seats.

That was, perhaps, the best summer of my life, with trips to Vancouver BC, Seattle WA, Oakland CA, Los Angeles CA, Calgary, Edmonton, Toronto, Winnipeg, and New York. My two navigator colleagues, Mr Jager and Frans Heuvingh, liked to be home in Amsterdam for their weekends, which meant I could have all the weekend long layovers on the west coast. That gave me time to rent cars and wander off up country to explore the scenery and seek out old aeroplanes to photograph. On one occasion it led to a rather unpleasant experience. I arrived back at the Hacienda Hotel at LA airport, late one afternoon, and some of the girls asked me if I would take them for a ride around Hollywood. I had the car until the next day, so I agreed. An hour later, we were pulled over at a police roadblock, whisked out of the car at gunpoint and told to 'assume da position'. We were frisked and treated with a lot of suspicion, probably due to our unfamiliar accents. Eventually, we convinced them we were a respectable Dutch airline crew, and we were allowed to go. Watching TV news next day, we realised we had been in the police dragnet looking for the 'Manson Family' after the Sharon Tate murders.

Other long breaks in Los Angeles gave me time to take all the examinations for an American Flight Navigator licence. That was somewhat easier than the British one and rather more practical, less theoretical. It still took two whole days, even though all the questions were multiple choice. In later years, I did all the tests for the American FAA Flight Dispatcher licence in case it might someday be useful to me. That is a licence required by all operations control officers working for major US airlines. As with all FAA exams, it was multiple choice, but also involved almost a whole day of verbal questioning about flight planning, meteorology, air carrier law and so on. In fact, over the years I also collected an American Air Transport Rating, while being checked out as a commercial seaplane pilot, as well as an Airframe and Powerplant Mechanic.

Long breaks in Vancouver were convenient too, enabling me to visit my aunt now and then. My six-year-old cousin was once overheard explaining to her school chums, 'This is my cousin David … he's an alligator'. My aunt is only a year older than I am, so I came in for some rather ribald jokes from the rest of the crews, after she'd dropped me off at the airport.

I found very pleasant digs with a family in the village of Oudekerke. They were quite happy about my comings and goings at odd times and only charged me for the days I was actually there, maybe two a week. If I had more than two days off in a row, the company were helpful in getting me rebate tickets back to Heathrow, on condition that I returned non later than the penultimate flight back to Amsterdam, the night before my next trip.

Work fell off at the end of August, though my contract was to the end of September. It was agreed that I could sit out the rest of the month at home so long as I was close to a phone. Then, at short notice, five cargo flights to San Juan, Puerto Rico came up. Those were all overnight each way, with a day stop in the airport Holiday Inn. Not a good spot for a day of resting as it was close to the end of the main runway. During the day, there were a lot of flights by Lockheed F104 Starfighters of the Puerto Rico Air National Guard, and each take off made the entire hotel jump about. On the very last service we were delayed a few hours on departure and left in daylight. This afforded me an hour or so wandering around the ramp taking pictures of many and varied exotic propeller airliners.

Martinair's long legged DC8-55 which could fly Amsterdam Los Angeles non-stop.

Chapter 11 - Tradewinds

There was no chance of any work with Martins over the winter, but the job market was good, and I was offered a permanent position back on the Britannia with African Safari Airways. During the summer, however, I had been in touch with Wally Phillips, a former British Eagle navigator who held the 'chair of navigation' with Tradewinds Limited, a new company operating six Canadair Forty-four swing tail cargo aircraft out of Gatwick airport. The firm traced its lineage back to Air Links Ltd, formed in 1959 with a single C47 Dakota. For some years, the company flew old BOAC Canadair Argonauts and the very last Handley Page Hermes in service. In 1965, a name change to Transglobe heralded the introduction of three Bristol Britannias to be used on inclusive tour holiday flights and long haul charters, especially over the North Atlantic. Transglobe pioneered polar flying and made history as the first UK airline approved to operate within the Arctic Circle.

In a 1967 lease/purchase agreement with Seaboard World airlines of America, new equipment arrived in the form of the first of six Canadair CL44D aircraft. Four had been handed over by November 1968 but the knock-on effect of other airline bankruptcies led to the collapse of Transglobe. On behalf of Seaboard, Transglobe had been operating a scheduled cargo service to South Korea and Japan, so Seaboard set up Tradewinds to keep the contract going. But in order to qualify for an Air Service Licence in the UK, the company was sold to Mr Charles Hughesdon, deputy chairman of London Insurance brokers Stewart Smith, who held the insurance account for Seaboard.

The CL44 was a development of the unbuilt Britannia 400 series, produced at Cartierville, Montreal. Engines were four Rolls-Royce Tyne 515 turboprops, a far more efficient power unit than the troublesome reverse-flow Bristol Proteus of the Britannia. It was a heavier aeroplane than the Brit, grossing out at a maximum weight of 210,000 lbs, (95,254 kg) but other improvements in detail design allowed a maximum payload of almost another ten tonnes.

Wally checked me out on a flight to Kingston, Jamaica and back. The return trip was empty, so we were able to take on full fuel tanks. That demonstrated the amazing range of the CL44 since we were able to do it without a stop, despite holding over Ibsley beacon for about one hour waiting for fog to clear at Gatwick. Even when we landed, we still had enough fuel to get to a dozen different diversion airports.

As in the Britannia, navigation equipment and methods were quite basic but at least the modern Edo 345 LORAN receiver was much better than the old wartime type APN-9 sets in the Brits. The periscopic sextant was the excellent Kollsman type, and the aircraft came with radar altimeters, allowing some use of pressure pattern navigation techniques.

A welcome feature of the Forty Four was a small galley and a rather cramped toilet on the flight deck. Cargo aircraft tend to be very cold inside, especially when loaded with fruit or vegetables, but we were able to keep warm and cosy on the flight deck, rarely needing to open the door. Somehow or other, the poor loadmaster would find a way of keeping warm. Usually, he would have come on duty some hours before us, so often spent much of the cruise wrapped up in three layers of sleeping bag. They were great guys and often made the tea or coffee or cooked quite good meals for us. However, I do remember the legendary Mick Miles, delivering a tray of drinks and announcing, 'That's one hundred cups I've made this trip. You can do the bloody rest yourselves!'

Mick was the senior loadmaster and a real character. A

pre-war veteran of the Short flying boat factory, he almost never wore a uniform. Instead, he would turn up in a lounge suit and would march straight through a security check in places like Libya, totally ignoring the heavily armed guards. When we pitched up at Gatwick early one morning with about fifteen tonnes of scrap silver from Dubai, a youthful customs officer peered into the hold and announced 'Oh, I've been told you have a load on board'. Mick replied 'Yup, yer lookin' at it – fifteen tonnes of silver'. 'Haven't you got any security guards then?' 'Nope, don't need 'em mate. Just try liftin' one of them bags'.

Many of my flights were to Africa where a popular destination was Asmara, Ethiopia, at an altitude of about 8,000 feet. That clearly limited available payload, which was always fresh vegetables, usually bell peppers. Take-offs were quite interesting in another way. The ground fell away rapidly after the runway end, so climbing through just a few hundred feet, and still sucking up the gear and flaps, we would immediately see mountain tops popping through the cloud deck below us. The charterer always gave each crewman a box of peppers which were much appreciated and graced dinner tables at home for many days after a trip.

Another popular destination was Lusaka, Zambia, which required a fuel stop at Entebbe on the way south. For some reason, on one flight, there was a misunderstanding about the diplomatic clearance so, until that was sorted out, we were placed under house arrest in the finest hotel in the city! Somehow, we managed to tough it out.

One of our captains, now sadly long gone, was somewhat addicted to the grape, to put it mildly. Now and then I would see his left hand lift a bottle from his flight bag so he could take a swig of his favourite tipple, Martini. It never seemed to affect his flying although he nearly did get us into a scrape on one occasion. We'd left Gatwick the evening before and dumped a load of cigarettes in Djibouti, a French colony on the Red Sea. After sunrise, we'd taken off for the short leg to Asmara. Tracking along the airway there was nothing much for me to do so I'd nodded off in a passenger seat right down the back. I was woken by the power reduction at the top of descent, and slowly we crept downwards. I felt the first stage of flap go down, followed by the gear. Then I felt the machine slow down as full flap began to run. Within seconds of touchdown the engines howled right up to full power, the nose rotating to climb attitude. As best I could I scrambled up to the flight deck and, as I opened the door, the first officer looked round and said 'Sorry, old chap. Wrong Airport'.

It seems there was another abandoned airfield about ten miles south of Asmara with approximately the same runway layout. Making a visual approach, it was quite easy to mistake them, especially when tired. More so I guess if you'd 'had a few!'. Thankfully, our copilot was more switched on that the rest of us, and realised the error when he spotted some local herdsmen driving a flock of goats down the runway.

Another interesting trip involved flying as a passenger to Beirut, to join a crew already there. The aeroplane was one of our US registered trio, with a rather multinational crew. I had a Pakistani captain, an American copilot, an Irish flight engineer and a Lebanese loadmaster. This left me as the sole 'Limey' contribution. I forget what the load was but we delivered it okay to Fort Lamy, now N'Djamena, in Chad.

A colleague reported for a flight and found that a very senior navigator from the UK Ministry of Aviation would be travelling as a supernumerary to Lagos, Nigeria, and back (probably just wanted to stock up with pineapples). As the flight progressed, my friend began to feel unwell, but he got them to Lagos on time. Then it became clear that Jim was quite poorly, with influenza or something, and quite incapable of working the sector back to Gatwick. So, our 'Man from the Ministry' was given an ultimatum: if he wanted to get his pineapples home, he would have to do some work for a change.

A serious setback at that time was getting involved in a fight with a burglar who attacked my girlfriend. He knifed me and got away. I lost a lot of blood and was off sick for almost a month. I guess that's what comes of using 'reasonable force'. If it ever happens again, I'll use any force I have. Never mind his human rights, as far as I'm concerned, burglars don't have any rights. He escaped over a garden wall, climbed a fire escape, and got into a large hotel through an emergency door. He must have had a smoke to calm himself because after he had walked out of the hotel and, probably, caught a bus, he left a dog end which set fire to a mattress. While I was being patched up in hospital, the hotel was on fire and being evacuated.

Chapter 12 - Zambia Airways

I stayed only six months with Tradewinds and joined Zambia Airways in early 1970. Flying the DC-8 seemed a better job and it was a three-year contract with possible renewal. All but two of the fleet were ex-British Eagle aircrew and we five navigators were sent to Alitalia for three weeks of training at Rome (Fiumicino) Airport. Mostly, that was a waste of time since we were all experienced on jets. Curiously, the navigators received absolutely no training in emergency evacuation or ditching drills, etc.

9J-ABR was a nice aeroplane with Rolls Royce Conway bypass engines. It was very well maintained by Alitalia but in every other respect the 'Alitalians' were not much help. Their operating standards were sometimes questionable, and it seemed that neither the Italian CAA, nor the Zambian DCA, had any real authority over senior Alitalia pilots, one of whom was seconded to Zambia Airways as Chief Pilot. We all got along well with the Zambian nationals with whom we came into contact, but it seemed to us quite unnecessary for Alitalia to be involved at all, except as an engineering contractor.

The work itself was somewhat routine, with only three scheduled services, and we only visited Rome, Nairobi, Entebbe, Lusaka and Athens. Duty times were rather long as well, not that I minded overnight flights, as that made

Flagship of Zambia Airways, our DC-8 having diverted to Malta due to early summer morning fog at Rome.

navigation easier. Three star astro fixes were much easier and reliable than daytime sun and moon fixes.

The Alitalia crew had always taken the airway route, the long way round, and were quite amazed that we were in the habit of taking direct routes right across the desert. They never seemed to use astro navigation, which we were pretty good at and used all the time. High altitude winds over most of Africa are not usually very strong and are fairly consistent, so we found that if we left the Ndola beacon in Zambia and flew a heading of around 340 degrees for about six hours, nine times out of ten it would take us right over the top of the beacon at Benghazi, Libya, and quite close to it the other time. That saved a lot of flying time, and we could usually do the Lusaka to Rome leg about forty minutes faster than Alitalia or BOAC. Of course, at times, we were almost two hundred miles west of the great circle track, but we made our radioed position reports giving our true position, so there was no danger involved. As we approached the Libyan coast, the stronger winds always blew us back towards the great circle track.

We flew one service on behalf of Alitalia which routed from Lusaka via Entebbe and Athens to Rome and then back a day later. One night, we enjoyed our usual coffee in the passenger terminal, then got airborne for the long leg up to Athens and breakfast. Unknown to us, twenty minutes after we left, a tank rolled down the hill and began shelling a large portrait of Milton Obote on the front of the terminal building. That was the night Idi Amin took over.

When we operated the southbound service a few days later the place had returned to normal, or so it seemed. The word was that the only fatalities had been two nuns and a priest killed by flying shrapnel. However, many years later, a book by the then East African Airways station manager claimed that the next morning there had been a complete bloodbath and he had even feared for his own life.

When the next southbound service was about to leave Athens, the station manager came rushing out to tell the captain there were fifty, new, last-minute passengers, and we needed to delay departure for twenty minutes. From out of the terminal building came fifty Ugandan soldiers in tiger suits and carrying guns. No way was the captain allowing that kit into the cabin, so the firearms all had to be checked as safe and placed in the hold.

All night long the captain received one message after another ordering him not to land at Entebbe, but to divert to Nairobi, and half an hour later, another order would reverse that. In the end, he landed at Entebbe where they all collected their 'artillery' and marched off happily. Not so happy was one passenger who was grabbed by soldiers and literally kicked to death on the ramp.

Three- and four-day layovers in Lusaka were a bit of a bore, despite having to tough it out in a five star hotel, so sometimes I would join an HS748 crew and ride the jump seat on local flights to airports like Mongu, way out in the bush. Arrival in Livingstone was interesting, giving a fabulous view of 'The smoke that thunders' Victoria Falls.

One time we had a lady copilot on loan from a British airline, who habitually wore a uniform skirt. I'll never forget the look of utter amazement on the faces of native loaders as she descended the aircraft steps. They didn't know how to address her, so just called her 'Sir'!

Gatwick-based Canford Aviation Services had an interest in a local air charter company. Now and then I would unofficially take urgently needed items like aircraft manuals for them as part of my personal baggage. We always arrived early morning so, after some rest, I took the company bus back for the twenty-mile ride to the airport. The driver dropped me off at the hangar and I delivered the goods and had a coffee with the engineers who offered to give me a lift back to the terminal, half a mile away. I said thanks, I was quite happy to walk, but the chief engineer warned me not to try taking a short cut through the vacant lot next door because a huge python lived in the long grass, and he'd already attacked two Africans.

Flying with the Chief Pilot between Entebbe and Athens one night, I was working from the copilot seat using my clipboard, with the two copilots asleep in the bunks. That was usual on long 'airways' sectors when I didn't need to use the sextant. After an hour or so, Roy announced he needed to use the toilet. Says he, 'You don't want me to wake those pair up do you?' I was somewhat dumbstruck that he would be happy to leave it all to me, a non-pilot with about ten hours dual training on Austers ten years before. Anyway, he quizzed me on what action to take in event of a yaw damper failure, pressurisation failure, runway stabilizer and an engine fire. I guess my replies were adequate for next thing I knew I was speeding through the African skies at Mach 0.82 with nobody else on the flight-deck but the flight engineer (asleep). Good job the passengers never knew.

About that time an uncle of mine was chief engineer of the Water Board in Kampala, Uganda. One day, one of his labourers failed to appear on time, but Bill didn't think that was unusual. Sometimes they would just disappear, then come back six months later and expect to carry on where they'd left off, and the first thing Bill would know about it would be a knife fight between the former employee and his replacement in full swing in the car park. Anyway, the next day the man pitches up for work with a sack over his shoulder and explains that he'd been attacked by a python that fell onto him from a tree. It coiled round him and was crushing him to death, so with his last reserves of energy he bit it over and over again. Somehow, he had actually killed it. Bill was having a hard time believing the story until the man emptied the sack and tried to sell him the skin.

The long duty times were not so bad for navigators since we had no real work to do on the short sectors such as LHR or Athens to Rome and with three pilots in each crew, all of them got at least some rest on every duty. The Flight Engineers had to be on the panel all the time though, so it was decided to have a second engineer on each crew. Unfortunately, I discovered that too late, otherwise I'd have easily cross trained as a flight engineer and stayed on.

One copilot was quite an unusual character. You don't find many former monks in the flying game, although actually he was the second one I've known! Nor do you find many pilots fluent in five or more languages. I asked him how he had somehow become a pilot and he explained: 'Well, Dave, you see, it was the Christmas party in the seminary in Montreal. Myself and one of the nuns were caught with our skirts up'. I believe he had a university degree in theology too. He came to the DC-8 fleet straight from flying single-engined De Havilland Beavers and eventually became Chief Pilot of the airline.

I noticed that one of our African cabin crew had some of her front teeth missing. I politely asked what had happened to them and she replied she'd lost them in a fight with another woman. Such fights, were, it seems, quite normal in her village.

We were all supposed to be there on a three-year contract, but few crews remained for that period, about half quitting within a year. Two pilots went to fly another DC-8 for 'Emperor' Bokassa, the dictator of the Central African Republic, and I was offered a job as navigator there. I was wise enough to decline, since I don't believe the job lasted very long. If anything, Bokassa turned out worse even than Idi Amin.

There's a nice little story about Johnny, the South African ex-squadron leader and complete nut case who'd offered me that job. One evening, we were waiting for transport from out hotel in Ostia, when he wandered across the street into a place where they made garden ornaments. He espied two huge stone falcons, or maybe they were eagles, but whatever they were, they were big. He bought the pair and cajoled us all into dragging them into the crew bus, then up the stairs into flight dispatch at the airport, where they remained until we came northbound a week later. We duly dragged the wretched things onto the crew bus and shoved them into the forward hold of our DC-8. Back at Heathrow the customs man couldn't find them in his book so let them through without any duty. Eventually, they were hauled into the crew bus and thence into Johnny's station wagon, last seen heading westward for Wokingham.

Days later, I was expecting to fly with Johnny on my next flight to Zambia, but instead the standby pilot turned up. When I asked the station manager where Johnny was, he casually replied 'Oh, he's off sick. Broke his ankle falling off a wall. Bloody fool was putting one of those damned chickens on his gate post.'

Another *faux pas* of his was having a few too many and celebrating by going round with a pair of scissors cutting everyone's neckties in half. The hotel manager did not appreciate having his latest Yves St Laurent creation thus ruined. One of the cabin crew got us thrown out and permanently barred from one hotel, but that story really is too disgusting to relate !

After about a year, during which I had somewhat reluctantly got married, I was offered a job on the Boeing 707 with British Midland. The work was much more varied and interesting, being mostly free gyro polar flights to Seattle. I accepted and gave my notice to Zambia Airways. This turned out to have been a mistake for the BMA job only lasted for two years. Also, soon after I put my notice in, I learnt about the second engineers being added to the crew. I would have done a lot better if I had asked to be cross trained as an engineer, since I could have easily passed all the written examinations (I later did so without any further studying whatsoever). Furthermore, I could have done that at almost no cost to the company and would have been assured of a job as long as they operated large jets. On the downside, I would probably have had to accept being based and paid in Zambia after the first three-year contract, but by then I would have been both DC-8 and B707 type rated and able to get a flying job almost anywhere.

British Midland Boeing 707 at Singapore in 1972. I thought were were going to have to stay there for some time because, in the process of doing a walk round inspection, my captain sustained an eyeful of hydraulic fluid.

Chapter 13 - British Midland

At first, I really enjoyed the BMA flying but all too soon the Seattle contract was lost and we resorted to ad hoc charter flying. This was still very varied, taking me to the Caribbean, and the Far East as well as the USA and Canada. The trouble was that I was getting even less time at home and my wife began complaining about my constant absence. My reaction was that I had to make the best of it while I could, since the job could not last for more than a few more years. I was enjoying the work and making good money so that we could look forward to having paid the mortgage off in a couple of years and be financially secure for the rest of our lives.

Penny, alas, came from a family of rather bad boozers and I think that she was also a manic-depressive. After one year of marriage, I learnt that she was playing around with one of her work colleagues. Almost exactly a year after we married, she left me and said she would file for divorce. By then, I was quite fed up with her attitude and was glad to see her go. I saw a solicitor who advised me that so long as I kept to the straight and narrow for a year or two, he would get me out of it at very small cost. In fact, he did very well for me, but that meant I had to keep out of any relationship until the decree absolute was issued. I lost out on the opportunity of getting into a relationship with any of the girls I was flying with, one of whom was making some obvious blandishments towards me. Sure enough, I fancied her, but in the end, she got fed up and married someone else.

The closest scrape I ever had was in the Boeing 707. I was navigating late one summer morning and we had just launched from runway 23 at Stansted. I was attending to the paperwork and not looking outside. Nor was anyone else, after all, we were in cloud on an IFR clearance within controlled airspace, under radar control and quite busy anyway. We had passed through about 4,000 feet and turned over Matching when we popped out into the sunlight. Then I felt a violent bump and saw check engineer Jim Hawkins, on the jump seat, instantly raise a hand to protect his face. I heard screams coming from the cabin. Jim and the pilots had a split-second glimpse, but I never saw the Piper Apache or Aztec twin that had just passed between the engine pods on the left wing.

I can't recall the outcome of the subsequent air miss investigation, but our pilots were certainly not at fault since they were looking almost directly into the Sun when we came into the clear air. Anyway, the good lookout and quick reactions of the Aztec pilot who, it turned out, was also adhering exactly to his air traffic control clearance, certainly saved 189 passengers and eleven flight crew, giving me another forty-four years so far. Good luck also that he shoved the control column forward and dived below us. Upwards and we would all have had it. But just imagine how he felt when a bloody great Boeing 707 suddenly reared up ahead of him at a closing speed of over 600 Knots. His laundry would have certainly appreciated the event's significance!

Arriving mid-afternoon in Toronto, as was my custom, I was on the jump seat behind the captain for the landing. We must have been as low as five hundred feet when Fred exclaimed 'There's something on the runway!' and called for immediate climb power for a go around. Then I saw it. Something green was moving in the 'piano keys' at the runway threshold. As they sucked up the gear and rotated into climb attitude, I glimpsed a Piper Cherokee trying to clear the runway via a high speed turn off, which was blocked by a Cessna Twin.

Eventually we got round the visual pattern and landed. In the office Fred called air traffic, said who he was, and asked for some explanation of why he had been cleared to land when they knew there was an aeroplane on the runway. I could hardly believe it when they replied that the Cherokee pilot was an Air Traffic Controller himself, so he knew what he was doing. If Fred hadn't spotted him, he'd probably have been squashed flat by well over a hundred tonnes of Boeing moving at about a hundred knots, while the lives of 189 passengers and ten crew would have been at risk too. I still take Fred for a drink now and then—he's a very fit ninety-four!

By then, navigation technology had reached the point where no one ever really got lost, perhaps 'a little uncertain of position' as the saying went. However, I nearly did put up one black. We were heading for Entebbe, but the Boeing had a recent history of autopilot problems. The company sent along an engineer to see if he could figure out the glitch. No sooner were we airborne than he announced that he was not an instrument fitter, but an airframe man. A fat lot of use he was when the autopilot quit yet again. With the pilots obliged to hand fly for hours, the accuracy of my sextant shots wasn't up to the usual standard, but I muddled my way across the Sahara Desert and the Sudan, until Murchison Falls came up in the right place on the radar. I was dog tired and nodded off to sleep in my seat, for it had been a long day. Unfortunately, the pilots didn't wake me up to say they couldn't receive the radio beacon somewhere near Gulu in Northern Uganda. We had been told that the VHF beacon at Entebbe was off the air for maintenance, so I had 'tweaked' the final magnetic track on the flight plan about five degrees to the east. If we couldn't get the other low power beacon at the airport, at least we would know it was to the west of us. This is known as the 'Landfall Method', which many years before had been invented by Sir Francis Chichester, the famed solo round the world yachtsman.

When I woke again, we were down to 6,000 feet and getting nothing from any of the radio beacons. The swamp coastline didn't give any useful radar returns but, fortunately, the night weather wasn't too bad. We were lucky that my skipper knew the area quite well and was convinced we were to the east of the airport. It turned out he was right, and he got us down in one piece, although I was very worried for a short time as we barely had enough fuel to divert to Nairobi.

There were always a few pranksters around, mostly among the cabin crew. On empty legs we normally put our suitcases in the passenger cabin, so we did that on an empty flight into Toronto. On the way over, some of the girls opened up the copilot's suitcase and neatly laid out a bra and a pair of panties on top of his kit. Along with a packet of condoms and a few other 'ladies requisites'.

Soon as we arrived, the first steward had nipped off into the terminal to tip off the customs, so they were in on the joke. They must have selected the guy who was best at keeping a straight face, so when we were all asked if we had anything to declare he made a point of selecting our copilot's suitcase for inspection. Well, you can imagine the look of the copilot's face when the contents of his bag were laid to view.

He was very quick witted though and soon summed up the situation, selected a Tampax from the packet, held it and sniffed it as if it were a fine cigar, then placed it behind one ear, announcing that he would smoke that later. One elderly lady customs officer I noticed seemed so embarrassed I thought she was going to pass out.

One winter night we had a two hour wait in Ottawa, and the best place to eat was the airport staff canteen. A stewardess approached the chow line and asked what was on the menu. 'Stoo, stoo, and stoo' announced the huge deadpan Canadian cook. Our girl soon caught on so asked 'What was that first one?

'Stoo'

'What was the second one please?'

'Stoo'

'Sorry, again, what was the third one?'

'Stoo.'

She pondered for a few moments then said,

'Is it possible I can have some stoo please?'

'Yeah, sure you can have stoo if you wannit.'

Off she trotted with her 'Stoo'.

Next was the captain, who went through the same procedure. By the time everyone had their 'stoo' served, we were nearly all in fits, but the cook was as deadpan as ever. Great people, Canadians.

We were in the process of applying to operate north of the Arctic circle, something I had been doing already for Martinair. The UK Rules, however, were somewhat stiffer, and we were still limited to low latitude routes which added about an hour to the Stansted/Seattle sector. I discussed it with one of the captains and he agreed we would take the short route. I had to keep two logs and two charts, so that we had fake ones for the authorities' benefit and real versions for us. Double the workload for me, but at least it saved the company a few bucks and the passengers were always happy to get there a bit faster. Fortunately, nobody in authority noticed and there were no embarrassing questions asked later.

The flight planning tables for the Boeing 707 were not very well designed or very legible. The B707 and the DC-8 had the same engines and the same maximum take-off weight. The main difference was about four degrees more wing sweep on the Boeing, so I tried using the Martinair DC-8 flight planning graphs for Boeing flight planning. It worked very well indeed, so I carried on doing that for years.

At intervals, flight crew are required to practice ditching drills, which include climbing into a dinghy from the water. Two senior stewards were placed in charge of organising it in Saffron Walden public baths early one summer morning before the schoolkids arrived. There were three flight deck crew and a course of a dozen or so new stewardesses. One of the stewards chucked the twenty-five-man dinghy into the bath and pulled the lanyard, whereupon there was a loud 'whoosh' and it inflated. Fine, except for one problem: in one direction, it was bigger than the bath!

It took quite a while to deflate and re-pack it, then we were able to have another go, this time using a smaller ten-man model. For realism, I always turned up for these events in a pair of overalls, but all our young ladies pitched up in their bikinis. We inflated the life vests and were happily splashing about in the water, waiting our turn to climb in, when one of the girls got her bra straps entangled in the dinghy ropes. Being such gentlemen, we naturally averted our gaze whilst they formed a protective circle so she could take her bra off and sort it all out.

Stewardesses are great girls—full of fun and some of their antics used to cause us much amusement. One appeared at a crew party—we'd all gathered in the skipper's room—in a terrible state of hysterics with the tears flowing. When we'd calmed her down a little, we asked her what the problem was. 'Someone was looking at me through the keyhole of the connecting door', she said. When asked how she knew, she replied 'I was looking through the keyhole the other way'.

We picked up some work flying cargo from Tel Aviv to Hong Kong. That meant flying the Bangkok to Hong Kong sector along the Green One airway over South Vietnam. The war was still going on and at night we could often see gunfire on the ground, although we were a hundred miles south of the demilitarised zone (DMZ). Near the east coast, we could see what seemed to be parachute flares reflected in the water and once, in daylight, even saw some B52 bombers heading north. Later, a few years after I had left the company, a BMA Boeing was one of the last aircraft to leave Saigon on a relief flight loaded with orphaned children.

Hong Kong was a fun place and very good for cheap cameras and such. But one night the copilot and I went out for a meal and despite the variety of dubious looking food offered by street traders, we couldn't find a Chinese restaurant. I think it was the Colonel's Chicken we ended up feasting on. Lots of American troops from Vietnam were on 'R & R' there at the time so there was a classic headline in one of the local newspapers: 'BROTHELS HELP OUT WITH HOTEL SHORTAGE'.

When we were not too busy, Chief Navigator Reg kindly turned a blind eye to some of my freelance activities between trips. One time I was loaned to the British Aircraft Corporation to assist with the re-delivery of a Viscount airliner from Bournemouth to the USA, on its way to Uruguay. Using an internal ferry tank, they had intended to fly the South Atlantic route, but the tank leaked so badly on a test flight that they had to leave the aircraft for days while the kerosene evaporated and dried out.

By the end of 1963, when the final Vickers Viscount rolled off the Bournemouth (Hurn) production line, a total of 444 type 700 and type 800s had been built, making it Britain's most successful large airliner. A decade later, some 250 were thought still to be in service worldwide, many having found eager buyers on the second-hand market. Several South American airlines had been enthusiastic buyers. On occasions, they were rotated back to the makers for major overhaul. This generated a small amount of interesting ferry work for freelance navigators because most Viscount-rated pilots had no experience of North Atlantic routes and procedures. Also, the aeroplanes themselves tended to be only equipped for short range 'on airways' navigation, and did not carry the navigation aids vital for long over water sectors.

The trip that came my way involved an aeroplane of PLUNA, destination Montevideo. The first language of the crew being Spanish, I went along as navigator/radio operator. A large chauffeur driven limousine appeared outside my flat at the appointed hour and I was whisked away, in great style, to the old Vickers headquarters at Weybridge aerodrome. The luxury didn't last very long though. I made the rest of the 100-mile journey to Bournemouth with a BAC commercial executive in his Morris Mini-Minor.

At Bournemouth, I was greeted by a former Battle of Britain Spitfire pilot, Captain David Glaser, head of Flight operations, and introduced to the crew headed by Captain Willi Hrdlika, a Czech. We discussed plans for the trip and decided on an early start for Prestwick, Scotland, next morning. Their original plan was to route via Washington (Dulles) airport, but I happened to know that the US international Transport Exposition, Expo 72, was in progress at Dulles, so my advice to use Baltimore was accepted. After a good dinner and a night's rest ,we filed flight plans through to Keflavik, Iceland, using the call sign 'PLUNA 101' which I figured rolled off the tongue better than the aircraft registration letters.

Navigation equipment was very basic, just one radio compass, two VHF navigation radios and search radar in the nose. The only addition was an old wartime LORAN A receiver on a special rack in the passenger cabin using a jump lead to the radio compass sense aerial in order to receive signals. One thing I missed badly was a sextant mount, which meant I could not take a periscopic sextant along. The crew were quite proficient at 'Aviation English' but not accustomed to the high-density radiotelephony on an early summer weekday morning over southern England. The R/T was soon delegated to me, sitting on the jump seat.

There were five of us on the operating crew, Willi being assisted by two relief captains and Señor Mendoza, our radio officer. Two senior management types from PLUNA were also along as passengers. Mendoza spoke no English at all, but was a Morse Code wizard, easily receiving continuous weather reports well in advance of our arrival at each airport and keeping in constant touch with Base Operations, in Montevideo. He was very formal too; each time he came to the cabin to give me a weather report he stood rigidly at attention and gave a solemn salute worthy of the Brigade of Guards.

Refuelling at Prestwick, we filed our flight plan for Keflavik and negotiated our clearance through Atlantic air space. Heading for Stornoway on the Isle of Lewis, we levelled at FL160 (16,000 feet). From Stornoway we climbed to FL180 (18,000 feet) out over the ocean and with some spectacular views of the Western Isles. With my

Sodrestromfiord, Greenland with Viscount CX-UQN in 1972.

chart on a clipboard—there being no proper navigation station—I kept an air plot of our position and made fixes at about twenty-minute intervals using the LORAN system. After about three hours, we began to receive the Vestmannaeyjar radio beacon on 375 Kilohertz right ahead of us. That enabled us to make a small heading correction which took us directly on to final approach at the airport. By then, we had conserved fuel by climbing to 24,000 feet. With no modern distance measuring equipment, working out the descent point could be a problem, but radar and visual fixes on the south coast took care of that.

Night stopping at Reykjavik's Borg Hotel, we fired up our four Rolls Royce Dart engines early next day, and headed for Sondrestromfiord, now known as Kangerlussuaq, on Greenland's west coast. I had flown over Narsassuaq, furtherr, south a number of times and didn't like the look of it, or it's weather reputation. So though 'Sondy' was further away and increased the distance on the next leg to Canada, the forecast was much better.

Leaving Iceland behind, I was back on the LORAN again until the spectacular east coast of Greenland showed up on the radar and then gave me a good visual fix. At Flight level 180 (18,000 feet) we were about half the altitude I usually crossed at, so the views were much better. But we soon needed even more air beneath us because safety heights in Greenland are quite high and even in the clearest condition the snow reflects so much sunlight it's impossible to focus your eyes on the surface. LORAN signals faded as we flew on inland but soon I was able to get a fix from a USAF radar site on the icecap.

To help with the radio, I rode the jump seat on the descent, even using some hand signals to the pilots at times, until we came out of some cloud. When we did emerge from it, we were in the fiord— happily far enough from the rock face which appeared off the left wingtip. Sondy isn't a suitable spot to hang around for long. If the temperature drops below minus thirty degrees Celsius (-22F) most turbine engines are unable to start. Tanking up with (very expensive) fuel, we fled to Goose Bay, Labrador as soon as we could. Good LORAN cover in the Davis Straight made navigation easy and soon I was getting bearings on powerful radio beacons in Canada. The coastline is not well endowed with landmarks and the ice pack made the coast quite difficult to reconcile with our charts. By the seventies, very few commercial flights used Goose Bay, so we were parked with mainly military visitors from the RCAF, RAF and US Navy. Plenty of daylight was still available, so we elected to press on to Baltimore.

This was to be our longest sector, though fuel reserves were no longer much of a problem with so many diversion airports available towards our destination. However, we did need to fly as high as possible to conserve fuel. Navigation fixing was by LORAN for the first two hours, then we began to pick up signals from the old-style low frequency radio ranges which were still in use in Canada. Progressively climbing we finally reached flight level 260, the limiting altitude for that Viscount, and we were soon able to track the usual VHF radio beacons all the way to Baltimore. From that point my main concern became fuel monitoring and radio calls.

There was one mildly amusing interchange with air traffic control soon after we entered US air space. The controller came up with a request for our aircraft registration, saying he wanted our 'N-number'. I replied, 'Charlie X-Ray Alpha Quebec November'. After a lengthy pause, ATC responded with 'say again', so I repeated the phonetics, and after some deliberation at the other end, I heard 'Negative. I need your N-number'. 'We do not have any N-number, we have a CX number, we are Uruguayan' replied I. For a while there was stunned silence and then, 'Did you say Paraguayan?' At the time I think both Paraguayan aircraft and Cuban aircraft were banned from US air space. Not wishing to have a couple of jet fighters on our wingtips, I quickly replied ' No, URUGUAY—Uniform Romeo Uniform Golf Uniform Alpha Yankee Alpha November'. Finally, he understood, and came back with 'Okay sir, that's affirmative.'

Because we didn't have a device called a radar transponder, we were not allowed to fly so high over the USA and had to descend earlier than planned. Early evening saw us descending over the hills of Pennsylvania and Maryland. The crew wanted me to go all the way with them so they could show me the sights of Montevideo. I guess there were some very interesting old aeroplanes down there and I would have loved to visit. They even offered to fix me a free ticket back home from there, but my contract and insurance only covered me as far as Baltimore and I was needed back home in my regular job.

Reluctantly, I bade them farewell and wished them luck as they sped off to their motel. Next day, they would leave for Jacksonville, Managua and onwards, using the flight plans I had drawn up for them. But for me' it was over to the Pan Am desk and a ticket on the next service to London. Quite a long day that one—and I don't think the very attractive young lady I found myself seated next to could have felt I was very convivial company, for I slept soundly almost all of the way. I just hope I didn't snore.

And that Viscount? She served PLUNA well for many more years until replaced by a Boeing 737. In 1982, she was sold to an owner in the USA and flown to Tucson, Arizona, where she was eventually cut up for scrap.

From then on, the intervals between trips grew longer and longer and by early 1973 I was getting rather worried about how much longer the job might last. I decided to request leave, some unpaid, so that I could do a commercial pilot conversion course. BMA did not receive the suggestion with any great degree of enthusiasm and not very long afterwards me and two other navigators were laid off. The company had secured a good long-term contract with Sudan Airways and would not need us on their services. I was given three months paid notice and told that I was free to take another job any time I could.

I managed to get some freelance work back with Martinair Holland and with IAS Cargo, back on the Britannia. This tided me over until I could start the commercial pilot's course which was to begin in September and provided some interesting flying, mostly around Africa.

Chapter 14 - Pilot Conversion

I started the flying course with CSE at Oxford on Piper Cherokee aircraft, but didn't really enjoy learning to fly very much, despite having done about twelve hours on Austers many years before. I had also hand flown Chipmunks, Britannias and B707s a little and had many hours flying Link Trainer simulators, but that didn't seem to help much. In some ways, it may even have slowed down my learning. However, when I got on to instrument flying, poor old Dick Marsh, my instructor, was quite amazed at how good at it I was. I had not told him I had many hours in the Link Trainer behind me. Oddly enough though, when faced with the Frasca 101 simulator, I found that very difficult and my progress was quite slow. Probably because I was self-trained and not in the habit of using the 'Selective Radial Scan' of the instrument panel, as taught nowadays.

The first time I landed a Cherokee I unintentionally gave Dick a nasty fright. As soon as the mainwheels were on the ground, I pushed the control yoke right forward, to keep the nosewheel on the runway. That's a pretty effective way of wrecking a small nosewheel aeroplane, but exactly how you land a Boeing 707. The difference is the Boeing has hydraulic 'lift spoilers' which pop up out of the wing and dump much of the wing lift. Those are activated by a weight switch on the nose gear system, but of course the simple Piper Cherokee has no such refinements. The correct technique is to pull gently back on the yoke so that drag from the stalled wings slows the aircraft down more rapidly. Otherwise, a pitch oscillation can start causing the propeller to hit the ground and shock load the engine.

At that time CSE were very busy with major contracts for Air Algerie and All Nippon Airlines. Those contracts were clearly having priority in allocation of aircraft, so that often I flew only once a day, sometimes not at all. As holder of a Flight Navigator licence, I was exempt from all the CPL/IR written tests other than aircraft technical, which I was permitted to self-study. In theory, this should have made completion of the course in twelve weeks possible, but the shortage of aircraft was slowing me down. Out of nineteen Cherokees on the CSE fleet, some days eleven were not serviceable. I was almost certainly taking on too much by commuting from Heston each day—something I was not supposed to do under CAA rules for approved course training. Then the Arab/Israeli war broke out causing the 1973 fuel crisis. Light aircraft flying on Sundays was banned, slowing me down even more, and to crown it all many airlines were laying pilots off. The job prospects for a pilot were going downhill fast.

CSE may have had their problems with aircraft serviceability, which was probably down to delays in the supply of spare parts and not entirely their fault. But I will say one thing in their favour. They certainly didn't turn out 'fair weather flyers' for the weather had to be pretty bad for flying to be cancelled, and we did cross countries and local flying in some pretty grim weather. One problem I found was that although the instructors were very experienced, they all came from the military and, at the time, not one of them had ever worked for an airline. With their service background, they were all accustomed to what is known as landing on 'QFE'. That means setting the altimeter so that it read zero altitude when you are on the ground. Some airlines possibly still do that, but the vast majority land on 'QNH', that is, with the altimeter set to read zero at sea level. It's a simple matter of having a moveable plastic 'bug' on the outside of the instrument bezel which you move and set to the airfield elevation, then you fly right down to that. It has to be that way because airports like Nairobi and Denver are a mile high, so landing at those you would never have time to re-set to 'QFE' (by the way, those are not acronyms, but code letter groups. It's possible that QFE stands for field elevation, but the cynics aver that QNH stands for Questionable Notation of Height). I tried to convince Dick over and over again, but he just couldn't accept it was possible to land on QNH.

There were five other navigators doing pilot conversion at the time, and I think the staff had a bit of a problem figuring us out. Dual cross-country flights with Dick were sometimes amusing. Dick did look like a typical farmer but had actually been a fighter pilot, so wasn't really accustomed to dealing with navigators. Quite often, he'd get a bit worried about where we were but that never bothered me. Somehow, I'd always had the ability for my short term memory to retain the image of the map or the ground so I didn't really need to compare them very often and I always had a pretty good idea where I was. Except for one duty instructor, they didn't work weekends, but I didn't want to waste a Saturday when the November weather was perfect. I got the duty instructor to okay my flight plan and set off on a trip much longer than we'd been doing up to then. On Monday morning old Dick was a bit peeved to find I'd clocked up another six hours of solo cross country time.

Then I was offered more freelance work by IAS Cargo, so some weeks I was away three or four days. I had signed on as unemployed at the start of the course but when the Department of Employment found I was doing the course (making an effort rather than sitting at home waiting for a telephone call to tell me that they had found me a job) they cut off my benefit. In the end, I got so brassed off with the whole thing I decided not to become a pilot after all and quit the course with about eighty pilot hours. Except for the long 'qualifying cross-country', at least I had done all I needed to get just a Private Pilot Licence.

Then I got a very lucky break. I had known Tony Deighton as a navigator at British Eagle. He had been laid off at the same time as me but had managed to get into

South African Airways. In four years with them he had learned to fly part time in South Africa but had decided to come back to the UK rather than rejoin SAA as a pilot. Tony was at Oxford while I was there, just to do the tests for his UK ALTP. He had to go back to navigating for a few years until I got him a navigating job leading to him being checked out as a Britannia copilot. Then he was offered two jobs at the same time. Tony rang me one night and offered me one of them, which led to almost a year on the B707 with Bahamas World Airways.

Chapter 15 - Bahamas World Airways

The Boeing operated between Frankfurt or Brussels and Nassau, stopping westbound at Gander and Freeport. It was very well paid and gave me lots of flying and plenty of time off. I wanted to get my UK PPL issued but still needed to do the qualifying cross-country flight. That was difficult in winter and if I did not do it before the end of March, the validity of my flying test would lapse. Oxford Air Training School was still short of aircraft, so I went to the Nassau Flying Club. CFI Chauncey Tynes checked me out in a Cessna 150 and that evening I planned my route from Nassau to Rock Sound and North Eleuthera. The next day there were strong winds and Chauncey didn't seem too keen for me to go. I had to do it that day for we were leaving for Frankfurt the next day and by the time I would be back I would have lost the validity of my flying test at Oxford. So, off I went, a non-swimmer with just a cork floatation vest, over about eighty miles of shark infested water.

I made it to Rock Sound okay, even offering to orbit a few miles down the cost to allow a Pan American B707 to overtake me and land first. What a gentleman that Pan Am skipper was—while I was refuelling, he walked over and personally thanked me. I got a receipt for my fuel and got the customs man to rubber stamp my flight log. At North Eleuthera I topped off the fuel and did the same again—

just in case the Board of Trade chaps back home didn't believe I'd been there. Can't be too careful.

I set off back to Nassau but after about an hour the weather was getting worse. The sea was getting rough, and the clouds were getting lower. There was no sign of land at the appointed time, and I was starting to get a little worried. Then, way out on my left wingtip, I spotted something that didn't look like sea or cloud. I swung round and headed straight for it and after a few minutes it showed to be the coastline of Nassau. I made it back to Oaks Field okay and put the aeroplane to bed. Maybe I'd made some navigational error or perhaps the magnetic compass was a bit out of plonk on some headings. The next day we were checking out of the hotel for the flight home, when I began to idly examine the bookstore in the foyer. For the first time I saw a copy of *The Bermuda Triangle* by Charles Berlitz.

At the first opportunity I was up to the Board of Trade flight crew licensing office, which in those days was at Shell Mex House, in the Strand. All I needed to do was retake the air law test, since the validity of the one I had passed had lapsed. A very helpful operations officer invigilated and marked my paper and told me I had passed before my paperwork went to one of the clerks. A few minutes later he came back and said he didn't think he could issue my licence, since the qualifying cross-country had not been done within the UK. I knew that this was complete nonsense, since a number of pilots I knew of had done it in the USA, so I pointed out that there was no legal requirement for it to be done in the UK in the pages of CAP53. He still wouldn't take that for an answer, so I insisted that he see an Operations Officer about it. A few minutes later he came back with my licence. It seems that the boss had given him a flea in his ear and told him to issue it and he was not too pleased at having to do so. NDW one—petty bureaucrat nil.

Soon afterwards I got one of my friends, a Goanese Flying Instructor named Joe De Cruz, to check me out on the Cessna 150 at Fairoaks, which he reluctantly did. I then flew a friend up to Sleap, Shropshire and came back myself. After that I never flew a light aircraft again for two years and for most of that time never intended to, until Joe took a job as an instructor at Halfpenny Green aerodrome near where I lived and talked me into taking it up again.

Our Boeing was a very long-range model, originally specially built for QANTAS round the world service. We could fly direct from Nassau to Frankfurt and still arrive with stacks of fuel. Later, it was returned to National Aircraft Leasing, so we soldiered on with two old Pan Am Boeings that were pretty rough. One was diverted onto a contract for a series of flights to Luanda, Angola from Lisbon. We expected passengers to turn up for all the sectors but there were none to meet us at Lisbon so, instead of refuelling at Sal, in the Cape Verde Islands, we made it direct to Luanda despite being banned from flying over any of the African countries. The northbound flights were all full, but southbound all empty, so it was obvious that we were flying evacuees not tourists. Things were hotting up a lot in Luanda City ahead of independence from Portugal and sometimes shots were fired. The hotel was good enough but could supply no food other than coffee and toast so finding somewhere to eat became a problem. I did eight round trips, always arriving and leaving at night but the last departure was delayed until daytime. Out of the mist appeared two old de Havilland Dragon Rapide airliners that had been abandoned to rot. I hadn't taken a camera on the trip, but fortunately the flight engineer loaned me his. Happily, one of the Dragons was saved by an air museum in Brazil, but the other one probably perished. In the end, we lost the last few flights of the contract. There were too many passenger complaints about our cabin crew being so rude. During one layover in Nassau, the famous straw market was burned down. The local newspaper reported that arson was suspected because the fire had started in five places!

The job with Bahamas World was, perhaps, too good to be true. It lasted until late 1974 when they suddenly went bust, owing us all a month's pay. I suppose it was inevitable as we carried very few passengers on the regular Atlantic flights. Years later, when I read the book about rogue financier Robert Vesco and the IOS swindle, I realised it had been a vast money laundering operation which probably involved the Mafia. No wonder we always seemed to be carrying bales of 'promotional material' or 'airline tickets' between Nassau and our European office.

Chapter 16 - Freelance Navigator

I came back to London and heard that South African Airways were looking for some navigators. I approached them and was told to attend for an interview. On the spot, I was offered the job, based in Johannesburg, with routes to New York, London, Hong Kong via Singapore, and Australia. Wonderful! They sent me for a South African civil aircrew medical with Dr Ian Perry, which I passed. Then I was told to report to a London Hotel for another medical with the company doctor. That was when the blow fell: he failed three out of four of us, including myself, on eyesight. Dr Perry was quite annoyed, but there was nothing he could do about it. SAA were insisting upon military standard aircrew medicals, even for navigators, or that is what we were told. In fact, I suspect it was political pressure from the SAA pilot union against employing too many British, although one of the three who failed was Canadian. I believe that at the same time they had tried to hire about thirty British pilots, and that was what had annoyed the SAA Pilot Union.

Fortunately, Henk Fransen and Jan Jager at Martinair came to the rescue and offered me another summer contract on the DC-8. That was not due to start until May, but I did manage to do a renewal line check on the Britannia with African Safari Cargo to Nairobi. When I got there, I found that instead of a quickie round trip, the aircraft was rescheduled and would be away for three weeks, operating mostly out of Damascus. IAS Cargo were phasing out their Britannias, but I went round to their Nairobi office to ask the East Africa Director, Robin Grant, if they had anything going back to the UK.

Happily, they had a US registered DC-8 heading for Gatwick next day. It was Inertial Navigation equipped but the crew, on loan from Capitol Airways, were not INS trained, so it still actually needed a navigator in order to take the faster direct route. Captain Jim Chandler was a great guy, and I guess that for most of the hours over Italy and France, I was in the left seat. Anyway, IAS Cargo were saved about an hour's DC-8 flying, and I got home for free.

British Midland also renewed me on the Boeing 707 with a flight to Lusaka and back, sub chartered to Zambia Airways. By May I was current on all three types as navigator: Boeing 707, Douglas DC-8, and Bristol Britannia.

As always, flying for Martinair was enormous fun, although some of the duty times were rather long compared to the UK. In those days, the Dutch RLD only seemed to apply hours limitations to pilots and flight engineers. Navigators and cabin crew just kept on going until we got there. I was often given a free hand as regards transit between US and Canadian airports, so was able to visit some air museums and air tanker bases on the west coast. One memorable road trip, this time with a whole crew, was from Edmonton to Vancouver. That took us nearly a week, in two cars. There were four of us plus the six cabin crew and the wife of the co-pilot, so we were very well looked after in terms of roadside barbecues almost every day. The weather was not very clear, but we certainly saw some fabulous scenery and I even managed a close-up picture of a grizzly bear—from the safety of the rental car.

Over the years I had developed an immense respect for the way Dutch aircrew operated and the quality of their aircraft maintenance, but there was one aspect I did not like at all, and it eventually contributed to a tragic accident. Unlike the British, Dutch Aviation Law was a lot more relaxed about having a navigator in the crew so that many sectors were flown by just two pilots and an engineer.

Martinair was part of an airline consortium known as KSSU, which stood for KLM, Swissair, SAS, and French UTA. A number of other DC-8 operating airlines were members and between them cooperated on engineering and operating procedures. For example, flightdeck layout was standardised, so that companies could easily loan aircraft to each other, and it was not unusual for a KLM London to Amsterdam service to be operated by a Venezuelan VIASA aircraft, with a KLM crew, just to use up a few flying hours before a major check by KLM engineering.

American Airlines had taken over Trans Caribbean Airways but did not require that firm's DC-8 fleet, so one of them eventually found its way to Martinair. It was bought up to KSSU standard but the few times I flew it we were still waiting for the search radar to be replaced by a system with 150-nautical mile range, instead of the 180-mile kit it already had. It also had an unreliable Doppler navigation computer that had defied all efforts to fix it.

The annual Hajj flights to Mecca gave Martins some work flying pilgrims from Djakarta, Indonesia to Jeddah, Saudi Arabia, with a fuel stop in Colombo, Sri Lanka. On the night in question, the crew comprised a captain hired quite recently from a Belgian firm that had closed, along with a first officer rather new to flying heavy jets. Neither had much, if any, previous route experience.

It is always a dangerous practice to base descent point calculation upon a radar fix, some airlines forbid it entirely since it's a procedure known to have caused a number of accidents. With no navigator on the crew, they had not made a reliable position fix for some hours, just relying on the Doppler radar computers. The probable conclusion of the accident investigators was that the crew had timed descent based on a radar fix of the eastern coastline, thinking they were about 150 miles from it, when in fact they were still 180 miles out. It would not have been possible to check against the Colombo airport (Distance

Measuring Equipment (DME) because that operates on line-of-sight VHF frequencies and would have been blocked by the mountain ranges. Throughout the descent the airliner was about ten thousand feet lower than it ought to have been at any point. Eventually, the crew radioed that they could see the airport lights through heavy rain, but it seems that was not the airport at all, but an electric power station up in the hills. Seconds later the DC-8 slammed into a mountainside at about four hundred knots. There were no survivors.

To me, that accident is still a matter of immense regret. The expense of using a navigator that night would have cost the company so little yet saved two hundred lives. I've flown that sector many times before and since the accident in Britannias and Boeings and I know for sure that the British procedure of taking a three-star astro fix at forty-minute intervals would have alerted the pilots to the discrepancy of thirty miles in distance to go. At least two thousand times British Eagle Britannias returning from Singapore trooping flights, or QANTAS immigrant charters, had approached Colombo from the east with no problems, and those aircraft didn't even have the luxury of Doppler radar.

Eventually, the DC-8 fleet was replaced by the much larger DC-10, the first one arriving in 1975. That was fitted with the Collins area navigation system and had no navigator position. There was a jump seat for supernumerary aircrew, so I was able to make a few trips to Spain and back on days I wasn't required to fly. All useful experience.

At the end of that summer, Reg Peake called me and offered a full-time slot back on the B707 at British Midland. I did a few trips but then the work petered out so they asked me if I would like to take over as Station Manager Kuala Lumpur, since we had two aircraft leased to Malaysian Airlines. That sounded great, but a problem with the law arose. Earlier that year I had stopped to avoid an accident on the M1 motorway. Another car hit mine and it escalated into a major pile up that blocked the M1 for hours I later became the star witness for the police. Because I was away so much, the hearings in Northampton Magistrate Court had to be postponed over and over again. In the end, the police threatened to subpoena me and arrest me next time I got back to the UK, unless I put in an appearance. BMA had to take me off the KL job and send me to Damascus instead, looking after the two Boeings leased to Syrian Arab Airlines.

The Magistrates' court was something of a farce. While I was giving evidence, one of the magistrates actually nodded off to sleep. Needless to say, a fast-talking lawyer got the guilty party off, scot-free, and the beak threw the book at everyone else but me. All the other drivers had their licences suspended, except for the young lady who caused it all. Even the MoT tester who signed off the car I stopped to avoid had his authorisation pulled.

I wrapped up my affairs in the UK and installed myself in the New Semiramis hotel in downtown Damascus, travelling out to the airport each day by taxi. That annoyed me a bit because there was a very nice modern motel at the airport, where my three ground engineers were staying. I kept asking to stay there, but the company insisted on me staying in town, with all the aircrew. That made it rather difficult to get any sleep because there seemed to be a crew party going on all the time, and usually in my room. All the toilets in the airport terminal were utterly disgusting so, if I needed the loo, I had to either cycle over to the motel or wait for one of our Boeings to come in. My transport was a Chinese manufactured push-bike, which did the job but seemed to attract a squad of aggressive pie-dogs that took great delight in chasing it. My predecessor had equipped the bike with an aerosol spray can full of ammonia. I became quite practised in allowing for wind drift and leaving a cloud of ammonia in my wake. That soon had the dogs rolling in the dust and coughing. Kindness to animals is very laudable, but nobody wants to get bitten by a rabid feral dog.

After a month or so I got brassed off with the long and erratic hours, the bad food causing constant gut trouble, and the fact that I really needed to be back to the UK, either looking for another navigation job, or getting on with pilot conversion. It was fortunate that I 'banged out' when I did, for soon after I left the ground engineers got into trouble over a smuggling racket, or something of that sort, and my replacement had to sort it all out and get them out of gaol.

Chapter 17 - African Cargo Airways

Back in the UK, my old flatmate, Mike Owen, had started his own airline, Redcoat Air Cargo. In practice, he didn't have any aeroplanes of his own but supplied flight crews to African Cargo Airways. That firm was registered in Kenya and had just one Britannia. It was a convenient way of avoiding all the hassle of having to apply to the UK CAA for an Air Operators Certificate. We just slipped some dodgy Kenyan Government official about ten pounds for every hour we flew. Shady business, but that's nothing new in the airline industry, and at least it was a job.

Mike had other ambitious plans and Redcoat ended up operating three Britannias of their own but, meanwhile, he needed another navigator and some help with various projects in the office. The first job I got was taking two crews out as passengers to Nairobi and shepherding them all through the process of getting East African licences. That meant a two-day crash course in aviation law convened in my hotel room at the Panafric Hotel. Happily, everyone passed the examination, so I shipped them off on the next East African Airways VC10 back to the UK and hung around for a few days before joining another of our crews to navigate the Britannia back to Rotterdam.

Eventually, Mike left the firm and set up Redcoat with his own aeroplanes ,taking with him some of the other crew, though I chose to remain. They operated successfully for ten years. One aircraft was lost in a fatal accident, but neither the crew nor the company were blamed. Their Britannias also flew all the flying sequences for a TV series called *The Buccaneer*, about an imaginary airline called 'Redair' and ran to about ten episodes. Watching it gave me great amusement since some of the storylines were based upon things that had actually happened, including things that had happened to me.

By then, I had sold my flat in West London and—between trips—headed back to my parent's hotel in Staffordshire. My means of transport was a BMC Mini which served me well for about twelve years, despite sometimes being abandoned in airport car parks, often for weeks at a time. The company operations office was, of all places, above a strip club just off Leicester Square. Fortunately, crewmembers were rarely required to attend there. My added responsibilities sometimes meant having to put in a day or to there, so I became an infrequent long-distance commuter. The train service from Stafford to Euston was pretty good and I could be in the office by ten, putting in nearly ten hours office work and then catching the last train north. Thankfully 'home' was just ten-minute walk from the station.

Most of our work was hauling cargo to or from Africa and the Middle East. Charter quotes were cheaper if cargo could be trucked as far as Manston, Kent, so that most flights started and ended there. Crews sometimes needed to report to other airports like Luton or Stansted and some flights even operated from Bournemouth or East Midlands. Even Ostend and Rotterdam were sometimes the closest the aircraft got to the UK, so we had to commute on the BAF Herald services from Southend, often carrying as much as eight or ten grand in cash or unsigned travellers' cheques for 'ships funds'. Many countries in the third world would not allow credit facilities for fuel, airport fees etc. That meant being sure to carry a Treasury Certificate authorising us to export large amounts of cash. One other airline crew forgot theirs and had an enormous sum of money confiscated.

Over two years flying with that firm, every trip seemed to be a drama. We didn't pay much attention to the rules of flight duty limitations either and some of the duty periods were thirty hours or more. The aircraft had three bunks down the back. In the cruise, there would always be one or two of the crew crashed out back there. So long as there was a pilot in one of the front seats, that was usually good enough. The other one would be occupied by the flight engineer, me, or even the loadmaster. In the end, we all became pretty good at operating from any seat although I never actually got to land a Britannia.

At that time I did not hold a flight engineer licence because the CAA wouldn't allow me to take the written examination. They claimed, correctly, that I didn't have enough practical experience. 'If you haven't done it before, you can't do it now.' East African registered aircraft had to carry navigators on all flights longer than 1,500 nautical miles, but on a leg such as Shannon to Cairo, I had nothing much to do other than the flight plan. At cruise level, the engineer would hit the bunk for a few hours, and I would work the panel and keep the log. Sometimes I could relieve one of the pilots or, using a clipboard, navigate from a front seat. That stood me in good stead when I did become a pilot, because it made line training very easy. In 1980, the CAA changed the rules, and I finally the flight engineer licence exams. I went in 'cold' having done no studying at all and got a pass with about 95% in all subjects. Unfortunately, by then, the trade was on the way out, so the licence was no use to me.

Airframe icing can be just as serious in the tropics as anywhere else. Heading south across Egypt one night, we were expecting some, but it began to accrete at an alarming rate. That slowed us down considerably and the pilots and engineer had a hard time getting rid of it. By then we had lost a lot of time and used up a lot of fuel but happily a sixty-degree left turn of the airway was coming up. The skipper and I reckoned we were out of radar cover, so I gave him a heading to cut the corner and, from then on, we were a good ten minutes ahead of our flight-plan. Nobody noticed.

Many people would think an engine fire could be the worst emergency a crew could experience. At least you have one or two shots in the fire extinguishing system to deal with that, but in my view a runway propeller is a lot more serious. We had launched from Kuwait with a load of apples that had been trucked from Lebanon. A couple of hours into the flight, the number three prop went into fine pitch and the engineer completely lost control of it. Very quickly it began to overspeed and we had to slow the aeroplane down. That meant reducing altitude or the aircraft would have stalled and maybe entered a spin. At the lower altitude, with the prop still running much too fast and the aircraft on the wrong side of the 'drag curve', the other engines had to work harder, so up went the total fuel flow. Riyadh was closed at night in those days, so we had to head back to Kuwait. Although we had taken off with enough fuel to reach even Khartoum and hold for two hours, we only just made it to Kuwait without very much fuel to spare.

Khartoum airport was a problem because there was no diversion airport big enough to accept a Britannia within a reasonable distance. The procedure was to arrive with enough fuel to hold for at least two hours. There was one case where a Britannia from another company arrived during an unexpected sandstorm. They held for as long as the fuel allowed but, in the end, just had to take a chance and make an approach. Thankfully, they got away with it, but for the final few minutes of that approach there must have been some very tense nerves on board.

Having no children, I always volunteered for flights over the Christmas holiday. Yet the only time I ever ended up away at Christmas was the year I spent the holiday in the Kuwait Hilton. We had a good Christmas dinner there, but booze was very expensive, even if you could get hold of any. I spent most of the stay on the hotel teleprinter talking to base operations about an engineering problem that had caused the delay.

On one flight to the Gulf, we developed a leaking fuel manifold in the number three engine. We shut it down and pressed on, opening it up again for the landing in case we needed to make a go around. There should have been a spare in the cargo hold, but wasn't, so we signalled base to send one out by scheduled airline. Back came the reply from the engineering manager, 'Try looking in the hold' before he went home for the weekend and left us to it.

This called for some improvisation, so the engineer took the defective part round to the airport maintenance workshops, since there was no proper aircraft engineering facility there. On the dump round the back, they found a scrap forklift truck with some hydraulic pipe that looked about right. One end was a right angle British Standard Pipe (BSP) fitting, but the pipe was a foot longer than needed, so it was just possible to fit it. That allowed us to ferry the aircraft to India for its next job, by which time the correct part somehow caught up with us.

Perhaps the worst experience at that time was developing a kidney stone. We were heading north over southern Egypt at night when I began to feel really ill. Cairo then was a 'Black Star' airport, and nobody wished to land there at night. Eventually, the captain gave me some pain killers and I passed out, coming round again as we approached Milan a few hours later. The pain had gone but I passed some gravel stuff in my urine. I hitched a lift back to Luton on a passing Britannia Airways Boeing and reported sick. An investigation found and removed three tiny but aggressive tumours from my bladder. Another lucky escape.

Another time we went to a place named Khamis Mushait—I think it means Thursday market—in Saudi Arabia. A generator had exploded in the power station, killing a couple of workers in the process. We had to collect a new generator from Rotterdam and deliver it. Weighing nearly eighteen tonnes, it had to be dismantled with the stator and rotor loaded separately. Easy enough in Rotterdam, but a different matter on arrival, where they had nothing more than one serviceable forklift truck. The Saudis got it unloaded in the end, but it took three days. With no available accommodation we had to sleep on the aeroplane while they were working and eat whatever was edible in the terminal.

Our commercial boys seemed to have a knack of accepting difficult and unsuitable loads. Maybe we got the work because all the competition were smart enough to refuse it. Oil drilling pipes were a good example: eight tonnes each and a bit too heavy to carry under your arm. Somehow those labourers got them loaded in Cairo, but I reckon they must have been directly descended from the blokes who built the pyramids.

In some places the loaders were a lazy lot and the only way to get any sense out of them was a few freebie handouts, usually cartons of cigarettes. I hit on the idea of giving them any unwanted clothes like old suit jackets, and that did the trick in some places. Maybe, to this day, some loader's grandson is still shuffling about Khartoum in my by now threadbare RAF greatcoat.

In those days the term 'agricultural spares' was a common mis-declaration for armaments or ammunition, though I doubt if it ever fooled anyone. It is, after all, not easy to disguise a couple of flak guns, and that's what confronted us on the first of five round trips from Damascus to Khartoum. Somehow, we got them there.

One fine day we were on our way north from Maputo, in Mozambique, where we'd dropped off a load of relief supplies. The aircraft was empty, and our next destination was Nairobi. I mentioned to the skipper that in twenty minutes we would be overflying the Ngoro Ngoro crater in Tanzania. We were way outside radar cover, so he dropped our 'Brit' down to about a thousand feet above the crater rim, and we did a nice circle right around it. Sod's law came into play though— I'd run out of film.

Fortunately, the overseas aid funded by the UK taxpayer doesn't all end up in Swiss bank accounts, but a lot gets frittered away on useless things that do nobody any good. We were contracted to fly eighteen tonnes of laboratory equipment to Mogadishu, Somalia. Two technicians accompanied us, one of whom explained to me that in crates down the back were six laboratory thermometers. They were so accurate that the Standards Division of the UK National Physical Laboratory only needed two of that quality. Until independence in 1960, Somalia had been two separate colonies, British and Italian. By 1976, in practice, they were still largely separate and not fully cooperating with each other.

We sped south-eastwards across French Somaliland and into Somali air space at about 18,000 feet in clear skies early one morning. Suddenly out of nowhere a Mig 19 appeared on our left and flew across in front of us. I felt the sudden bump of his slipstream, so he was pretty close. The periscope sextant was in place, so I lined it up to try and watch our six o'clock. Fortunately, there was no sign of any other Mig, but then a voice came on the VHF saying we had been 'intercepted by the military power' and must immediately land at Hargeisha. This was within the region of old British rule, but the airport runway was not adequate for an aeroplane the size of a Britannia. Unfortunately, we were not in any position to argue, so our skipper made a respectable landing on the short runway. As we taxied to the ramp, the Mig landed. It immediately burst a tyre and its undercarriage collapsed, then it slithered across the bundu and came to a halt in a cloud of dust. When the visibility cleared, we could see the hapless pilot running for his life. I imagine the Somali Air Force was grounded for the day. Anyway, our captain had made four kills as a World War Two fighter pilot, so perhaps this was the fifth that qualified him as an ace?

The engineer shut down three engines and kept number three running to give enough hydraulic power for me to open the cargo door. That meant I could not hear the anguished yells of Africans on the ground who were gesticulating frantically. It turned out they were worried that if we shut down the engines, we would be unable to start them again because the Air Force had pulled the same stunt on a Canadair Forty-four the previous month which had blocked their parking area for days. It needed an air start unit, which they didn't have. Fortunately, it was possible to do internal electric starts with the Bristol Proteus. One of its advantages, although it had always been a problematic type of engine many other ways.

Meanwhile, the airport authorities had been in contact with some potentate or other and established that we did have diplomatic clearance and we were free to go. Because Hargeisha airport was inadequate for a Britannia, we had no performance tables or graphs to figure out the correct speeds, although any fool could see the runway wasn't long enough at our weight and the prevailing temperature. It was now a matter of guesswork and eyeballing it. Somehow, we got airborne, clearing the boundary by not very many feet and staggering away in the hot, thin, air. If we'd lost an engine on the take-off roll, it surely would have been the end of the aeroplane and probably the lot of us, while six incredibly valuable thermometers would have been smeared across the desert.

To cut down on expenses at fuel stops it was usual to carry an aluminium ladder in the aircraft, so we could get in and out without having to hire air stairs. I guess the loadmaster had been having a bad day when we arrived late at Lusaka, Zambia, because we found he'd left it buried under the load. By the time we'd completed all the after-flight checks, the airport had closed, and everyone had disappeared. Calls to the control tower got no response, so there was only one thing for it. Muggins shinned down an escape rope and wandered off into darkness, eventually finding someone who could rescue the rest of us.

The Bristol Proteus was a free turbine: the propeller was driven by the final stage turbine and not directly connected to the rest of the engine. Parking brakes were fitted to the props so that they wouldn't freewheel in the wind, and it was standard practice to line up the four bladed propellers, as it looked a lot neater when the aircraft was parked. Sometimes a brake didn't work, and the prop would have to be secured in some other way, or just left slowly rotating. Not really a problem, except that if there was any tailwind on startup, the prop would be rotating the wrong way and, if the wind was more than ten knots, someone had to stop the huge sixteen-foot diameter propeller with his hands, then hold it steady whilst the pilots and engineer started the engine. Usually that honour fell to me, but although it looked a bit scary to an onlooker, there was really no danger to it at all.

The other problem with the curious 'reverse flow' design of the Proteus was icing. Years before, it had caused all four engines of a Britannia to 'flame out', although they had been restarted and the aircraft landed safely. Curing the problem caused major delays with the type's introduction to service and cost the Bristol Aircraft Company many lost orders. By the time the problem was sorted out, faster first-generation pure jet airliners were starting to eclipse the propeller-turbine designs. The Britannia airframe itself was a fine design. Although several of the eighty-one built were lost in serious crashes, none of those were attributed to any fault in the airframe. Perhaps another reason for its lack of commercial success was the complexity of the aircraft systems, especially the electrics. The view from the pilots' seats was not very good either and when the Canadians redesigned it as the Canadair Forty-Four, the flight deck windows were completely redesigned.

A frequent job was hauling meat, described as 'Bullock Mutton', from Bombay (Mumbai) up to Kuwait. We sometimes suspected that someone had been going round the city knocking off holy cows the night before. On one trip, we just happened to follow the track of a total eclipse. It got so dark that I was able to make a three-star astro fix in broad daylight, I doubt if that's been done many times. Invariably, on arrival at Kuwait, we were greeted by a reception committee of around a billion flies. In bright sunlight you could barely see our aeroplane for the black cloud that enveloped it and, of course, the refrigerated truck was always late meeting us.

One morning I thought I was in dire trouble. I swung round a corner in a corridor, on my way to the met office, and tripped over an Arab in the middle of morning prayers, sending the pair of us sprawling on the deck. Thankfully, he was very nice about it, waving my profuse apologies aside and inviting me to listen. I've never been religious or into poetry but sometimes it really does sound quite beautiful if you listen to their invocations.

Bahrein was another destination, and it normally meant a load of vegetables. By the time the aeroplane was unloaded, there was usually enough food littered around the ramp to feed an Indian village for a week. At least, that's how our Indian loadmaster described it to me, almost in tears. We briefly had a contract hauling tea from Entebbe to Khartoum, with a back load of laundry soap the other way. I was asked to go to the Nairobi office with some funds and meet the captain arriving in Entebbe that night. With about $10,000 US in unsigned traveller checks in my bag, I hopped an East African Airways flight and sat in the new terminal building for a few hours awaiting the Britannia. Apart from some Africans playing cards a few yards away, the place was deserted. Then I heard an aeroplane approach and the familiar clunk of the propellers going into ground fine pitch, so I knew it was ours.

The aircraft parked outside the passenger terminal to unload. I handed the cash over to the skipper and settled down in the nav seat. We ferried empty to Nairobi. Then, after a short rest, took a load of exotic flowers up to Germany. I had a spot of leave coming up and I had fixed to do a freelance trip to Vancouver, with Martinair. The return flight was to be from Los Angeles a week later, so I picked up a rental car and headed down the coast. Somewhere near Florence, Oregon, I had the car radio on and learned for the first time of the Air France Airbus hijack and the Israeli Raid on Entebbe. I had missed being in the middle of it by less than a week. Lots of things seemed to happen to me at Entebbe, and not all of them good.

Later, when I read the book about the raid on Entebbe, I learnt that the Israeli planners had been desperate to find out if the terminal doors opened inwards or outwards. Nobody could remember, even though the Israelis had been there training the Ugandan Air Force for Milton Obote. A shame they couldn't have asked me! I can still remember those plate glass doors I used to sit behind, sipping my coffee. Even the huge teak door handles are fixed in my mind, as are the overflowing ash trays. They opened inwards.

Cattle flights were quite popular because we were paid extra for carrying live animals or hazardous cargo. Most popular of all with the loadmaster, because all the way home he could be seen shovelling the droppings into huge plastic bags, which he carted off and sold down at the local gardening club. I don't think I would have cared to reside within miles of his backyard on a hot summer day.

That loadmaster was a strange character too. He'd been a squadron leader in the RAF Regiment but had been court martialled and cashiered for punching an airman. That gave him a chip on his shoulder, especially if he discovered the rest of the crew had been lower ranked that him or hadn't ever been in the service. He took a real dislike to one captain and an argument ended up in a fist fight in a hotel room, after which he disappeared into the night. Eventually, he must have concluded that wandering round Damascus at night wasn't good for his longevity prospects so quietly crept back to the hotel. Unfortunately, he had dislocated one of the skipper's thumbs and we had an unserviceable auto pilot. So that meant this navigator, with nothing more than a private pilot licence only valid for light single engined aircraft, had to do his turn flying the Britannia for hours.

One time we launched from Milan at night with a load of sheep, destined for Sana'a, Yemen. If you're going to lose an engine, there's always a good chance it will quit on the first power reduction after take-off, and that night was no exception. Not usually a problem, except that we were way above maximum landing weight. Milan is in a valley so we had to climb on just three engines in the holding patten and could not dump fuel in case we flew back though our own cloud of vapour. In fact, loss of one engine was not a major setback, as we had enough fuel to get there on three.

Rules for carrying animals were quite strictly enforced and if we had landed back at Milan, all the sheep would have had to come off and be given twenty-four hours in a lairage before we could reload them, even if the engine problem could have been easily fixed. Therefore, it made more sense to press on. Fortunately, it was a daytime arrival at Sana'a and we did a three engine ferry home for an engine change.

Another arrival at Sana'a was at night with nearby thunderstorms. Immediately after we landed a lightning strike took out all the airport lighting and of course, the standby generator failed to come online. Seconds earlier, and I doubt that we would have survived a three engine go around with landing flap and all those mountains around there, not to mention the field elevation a few thousand feet up, and hot with it.

Juba, in the Sudan, was another lovely spot the tourists haven't got to yet. Relief supplies for a famine were the payload and we were scheduled to arrive just as the airport opened in the morning. That made it easier to find because I could get enough three star fixes on the way across Libya during the night. About half an hour out we called Juba, but had no reply. A few more tries bought forth no response and we were getting very close to the descent point. The skipper 'transmitted blind' that we would descend to 10,000 feet. There was still no reply to the call as we approached ten and said we were carrying on down to 6,000. Same again as we approached 6,000, with the field in sight, at which point we said we would go down to pattern altitude (circuit height). We flew along the runway at a thousand feet, followed the visual circuit pattern and made another fly-by at five hundred. There was still no response to any of the radio calls so, as the runway appeared to be clear, gear and flaps were selected on the downwind leg. The landing was uneventful, and we taxied towards what appeared to be the terminal and shut down the engines. The next thing we saw was a chap approaching at high speed on a bicycle. Imagining we were in deep trouble for landing without clearance, we were relieved when he yelled 'Sorry boss – we've lost the keys to air traffic control.'

Not long after that, the operations manager wanted to loan me to a competing airline owned by an Indian businessman. They had two ex RAF Britannias operating as private aircraft on the excuse that he also owned all the payloads. This all seemed very dodgy as I knew for a fact they had been operating sub-charters for Air India and that was clearly unlawful. On the morning he had been expecting me to turn up at Heathrow Airport, I had refused to get involved. When told him I wasn't coming, he said he'd punch me on the nose if ever he met me. Whereupon the flight engineer observed, 'Not if I was you mate, Welch is a bloody big lad!'

Some years later, he arrived at Manston Airport in one of his aeroplanes. It was due for a maintenance check the next day, so the engineers put it to bed and locked up the hangar. As they were about to leave, he came rushing back saying he'd forgotten something and went into the aircraft emerging with a large parcel. The engineers were suspicious and phoned the local customs man at home. Some hours later, as the owner's Mercedes swept into his front drive, a squad of police cars surrounded him, and he was arrested. The parcel was full of some controlled substance and he ended up in gaol with his airline closed down. My brother-in-law had the job of cutting one of their aircraft up for scrap.

East African Airways was a fine airline operationally, and one of the few operators of the excellent British-built Super VC10 airliners. Sadly, infighting between the three East African states that owned it made its demise inevitable. When it finally happened, BMA was ready and waiting with the capacity to help Kenya Airways get up and running. I was still doing the odd freelance trip for them and was asked to ferry an empty Boeing to Mombasa to initiate the first northbound passenger service. The layover, in a beach hut motel, was very pleasant indeed, but things went badly wrong as we taxied out for homeward departure. The airport suffered a complete electrical power cut, and all the radios went dead. We sat at the runway holding point, burning valuable fuel for about twenty minutes, unable to look for light signals from the control tower. Then the captain had an inspiration. He got through to BMA Operations Control, near Derby, using the single side-band radio via the station in Enkoping, Sweden. BMA made an international telephone call to Mombasa Airport and miraculously got through in minutes. Thus was our take-off clearance relayed to us, and we were on our way back to Heathrow.

Eventually, African Cargo Airlines ground to a halt but my final trip was the usual drama. We had contracted to take some cattle, including a massive breeding bull, to Salalah, in Trucial Oman. The plan was to refuel once in Belgrade, and again in Muscat. Muscat was an awful place for a rest stop, and the company wanted us to do the whole trip to Salalah in one duty period, claiming that no hotels were available in Belgrade. That was clearly unlawful and there was a lot of aggravation before we left, including the boss threatening to close the firm down if we refused to go. Finally, we set off, but a few hours later, as we passed over Austria, we made a radio call saying we had shut down one engine and needed to land at Zagreb. On landing we shut down the remaining engines, then the me and the engineer performed a virtuoso bit of acting. We both spent some time peering into the vitals of the number two engine and discussing the fault with the skipper. He then announced that the engine had to cool before we could work on adjusting it. The Zagreb authorities seemed to accept this, so we ended up in a posh suite at the top of the best hotel in Zagreb. Mind you, I did have to sleep on a settee in the sitting room. Meanwhile, our prize bull chomped away on his food and seemed quite unperturbed by the proceedings. He behaved himself impeccably for the rest of the trip. Hopefully, his grandchildren and beyond flourish to this day out in Oman.

As there were no hotels there, we were billeted in some bungalows in a walled cantonment owned by Taylor Woodrow. The gate guard was a fearsome looking tribesman with a rifle about six feet long. When I finally awoke, the other guys in the same bungalow invited me to join them for dinner. One of them had overseen a road gang working on the M11 near Stansted Airport years before. He asked if I knew of a chap who had worked for him as a labourer but claimed to have been an airline pilot which, of course, he had some difficulty believing. In fact,

I did know of this guy, who was a bit of a character and at the time was, I believe, flying as a DC-10 copilot for Laker Airways. He'd lost his job when Lloyd International Airlines had gone broke in 1972. While he was looking for another job, he took some temporary work on a road gang. If ever I should meet him again, I was told to warn him that 'Big Paddy is looking for him!' Oh dear!

Chapter 18 - Instructing

Although we were busy and did a lot of flying, I still got quite a lot of time off at base. While on leave in the USA I had collected an American Commercial Pilot Licence (CPL), with Instrument Rating. The plan was to get a Kenya CPL issued on the strength of it, then get a Britannia type rating, and switch seats to copilot. Unfortunately, by the end of 1977 when things ground to a complete halt, that clearly wasn't going to happen.

During the winter of 1976/77 I somehow managed to attend a night school course in navigation science and technology. My fellow students included three very experienced ex RAF navigators and the others were all in associated professions, although I was the only one still actually still flying as a navigator. The course was absolutely fascinating because it also covered marine navigation, surveying, and space navigation. Some of the mathematics was rather challenging and I guess I must have missed about a third of the lectures due to being away flying, but I somehow managed to pass the end of course exams and came away with a very handsome certificate. Later, I was offered a place on a course for an MSc in navigation science, but I did not have time to take it up due to work pressures. By that time I was, anyway, into my fifties—rather late for a university degree to be much use to me.

Meanwhile, a friend had acquired the right to give instructor rating courses and was looking for a 'victim'. He offered me the course and I passed the test on the second attempt during that winter. Fortunately, he was also able to fix me up with a job as a flying instructor. Pay was appalling but at least I could take time off for freelance navigator trips, and the main thing was to clock up enough pilot hours to get my British Commercial Licence. For five months I worked without taking a day off, except for freelance navigator trips, until I had enough hours.

That provided a few tense moments as well. My first student was probably my best ever and I guess nobody could have been easier to teach. He was a former private soldier, fresh from a few tours in Northern Ireland. The worst problem was getting him to relax—at first, he used to sit to attention when he was flying. Anyway, he worked incredibly hard and passed everything with flying colours at the first attempt, eventually qualifying as a full instructor even before I did. We used Piper Cherokees as trainers but when he got his licence, he wanted a checkout on the Cessna 150, which was cheaper to rent.

My boss, CFI Wg/Cdr Mike Edwards was a real old-time instructor who insisted that checkout should include spin recovery. That almost killed the pair of us because, on about the third recovery at about 3,000 feet, something fouled the elevator control. It took almost all my strength to push the yoke far enough forward to stop the aircraft climbing, losing speed and stalling. We made a Mayday call on the airfield frequency, which didn't help any because the controller and the fire crew had all gone down to the village pub for lunch. Fortunately, another of our aeroplanes was flying and heard us, so at least I was able to tell him what our problem was. Not that he could be any immediate help, of course. We were not very from from the field, so I somehow set up an approach, though not on the duty runway. Then, at about a hundred feet, something suddenly went TWANG and normal control was restored.

Later investigation showed that, at some stage in the past, a bolt at the other end of the torque tube had been replaced with one that was about two inches longer than it ought to have been. It had fouled on a loose plastic tube which was part of the air data system while we were throwing the aircraft about spinning. It usually takes two things to cause an accident, and it was both the incorrectly secured tubing and the incorrect bolt that had nearly killed us. On that occasion, happily, it only caused an incident, but when that tube broke it almost caused me to nosedive into the deck.

I'm pleased to say that my soldier student is still flying, as a long-haul Airbus captain. Over the years, I did nearly three thousand hours of dual instruction and although I'd never really planned to do it in the first place, I really did enjoy it, especially the very basic training. I got a lot of satisfaction from taking on the students that the other instructors regarded as hopeless. I had found learning to fly quite difficult myself and I guess that gave me a certain empathy with other guys who were finding it difficult. Over the years, I have known other pilots who clearly had exceptional ability themselves but clearly didn't have the patience to make very good quality instructors. The same applies to some others who just thought they had exceptional ability!

Any aircraft cockpit is about the worst environment there is for teaching, so a thorough preflight briefing is essential. That can be boring for the student and cause lapses of attention. It's really a matter of striking the right balance and not making the briefings too long.

I had some very strange students at times. One had a pathological fear of dentists and a mouth full of rotten teeth, so bad that no doctor would give him even a third-class medical certificate. He had about four hundred hours of dual instruction and could fly like a professional. He was happy to keep on spending money though no amount of persuasion could drag him anywhere near a dentist. Another was a wings-qualified RAF pilot I had to check out in a light aircraft. He was fine handling the controls himself but the instant he handed over to me, up came his breakfast. That's not so unusual as it may seem, for I once knew an airline captain who just could not bring himself to look outside the aircraft unless he was strapped into the pilot seat. I guess I was lucky that I only ever had three that were airsick on me. You can guess who had the job of

Ace flying instructor 1983.

cleaning it all up. Fortunately, I guess I have a pretty strong stomach so that was never so much of a problem. My advice is always eat raspberry jam before you fly. Doesn't stop it, but the puke tastes a lot better coming back up!

Others had a problem because they wouldn't stop talking so just didn't listen. Some were fine enough pilots but had absolutely no sense of direction at all. One caused me a problem because he was so good, I really needed to send him off solo. Then, when I did, he got himself lost in the aerodrome circuit, in perfect weather conditions. Another got lost halfway around a solo cross-country exercise. When I sorted it out and got him back to terra firma, I found he'd left his navigation chart in his car, and he'd tried to do the trip without it. One lady student never had any navigation problems. Shona had spent many holidays and weekends canal cruising all over England. Each time we crossed any canal she could instantly recognise some canal side pub, or a lock system.

One of the 'talkers' nearly killed the pair of us. He already had his licence and was giving me a lift to another airfield in a Bellanca Citabria aeroplane. In the back seat, the only controls I could reach were the stick and rudder. Something caused the engine to quit at about eight hundred feet on final approach and he panicked. It was too late to try and restart the engine, so he was trying to do what all pilots are taught never to do: stretch the glide to the runway. There was absolutely no chance we could make it and there were two Cotswold stone walls on the way. Happily, I was much stronger than him. I managed to overpower him and land the aeroplane in a field of barley with no damage. I was yelling at him to put the flaps down for the last few seconds, but he was too panicked to even do that.

One or two former glider pilots were a problem because they, reasonably I suppose, thought they knew it all already. The problem was that not being used to the propeller wash over the rudder and elevators, they tended to be too rough on those controls at first. When starting a descent, they'd just shove the nose down and build up speed: not what you do in a powered light aeroplane. The correct way is to reduce power and maintain attitude to bleed off speed, then lower the nose and re-trim. We would always get them away solo though and, of course, in less time than a completely raw student.

One student pilot who was a farmer, ploughed right

A Tiger Moth I used to fly in 1983.

through an RAF zone ignoring all the ATC instructions and nearly hit a Hercules. They sent an Air Traffic Control Officer over to see me about it. When I interviewed the miscreant later, I found that the morning before he flew, he'd physically loaded a thousand hay bales onto trucks. No wonder he was half asleep in the aeroplane. Fortunately for him the RAF appreciate that students inevitably make a few mistakes, so no action was taken against him.

Another strange thing is that some pilots, even professionals, seem to have a mental block against some subjects. OI knew one chap who had enormous problems trying to understand the workings of piston engines and the CAA technical examination was a real problem to him. Yet he was clearly an above average operator, achieving a jet airliner command at a very early age. Many seemed to have an issue with navigation plotting. I quite enjoyed all of that, since it is such an interesting application of mathematics and logic.

I don't suppose it's any different now, but in those days there certainly was a lot of dodgy maintenance of light aeroplanes. I was once asked to take a very nice Piper Aztec to demonstrate it to a potential buyer. All went well until we took off to return to base. As I sucked up the gear, something went twang, and I could feel I had lost pitch trim control. Nobody has enough strength to overcome the stick loads on an aeroplane of that size, but by lowering the gear again and reducing the power, with the help of the passenger in the other seat pushing on the yoke, I managed to nurse the aeroplane back onto the ground in one piece. Poor chap had never been in a light aeroplane before, so I guess he probably never went in one again. It turned out that a trim cable had fallen off a pulley. Clearly the pulley guard had not been set up right but what really infuriated me was when an engineer casually remarked 'It's done that before—didn't you know about it?'

Some years later, air testing another Aztec I actually had a double engine failure. We needed to do a test climb on one engine, which went off fine. We levelled off at about five thousand feet and as I was about to restart the other engine, the good one decided it had had enough and quit. Luckily, we were about half way between two aerodromes and in fine weather. We alerted the fire crew at base but by the time we had lost a few hundred feet I had the other engine running and made a single engine landing, although I had to hand pump the gear and the flaps down. I guess I could go on for pages on the subject of shoddy maintenance. Two cases of asymmetric flap in the air, one of them at night. Gear refusing to lock down or refusing to come up, usually from dirt causing micro switches to malfunction and giver wrong indications. Part of the problem is that CAA engineering inspectors can't be everywhere all the time.

Worst of all was some of the dishonesty involved. I had one nose wheel collapse because a bolt had not been split pinned and fell out. The chief engineer simply put a new one in and swore it had been there all the time. He was trying to blame me, and the insurance company loss adjuster wanted to sue me for negligence. In fact, it never got to court because of the broadside the barrister hired by BALPA sent them. Perhaps the engineer realised that one of his staff was a friend of mine and would spill the beans if it ever got to court.

Another time I had a rough running engine halfway to an airfield where I was taking a small Cessna for a routine check. The chief engineer couldn't figure out what was wrong, so just signed the aircraft off and told me to take it. On take off it started to run roughly again, so I did a 'one-eighty' and landed back on the same runway. He said he hadn't been able to change the air cleaner element because they didn't have a spare in stores. Then, to my utter amazement and disbelief, I saw him take an old one out of a dustbin and fix it to the aircraft. I said I was not flying it without him in the other seat—so I spent the rest of the day getting back to base by road— no way could I get him into the Cessna. Later, I learned that the slow running jet in the carburettor was completely burnt out, probably due to a recent intake fire I had never been told about. I still can't figure why it didn't run rough all the time not just occasionally.

Some of the freelance navigator trips I did were for Transasian Airlines, with a Boeing 707. Mostly they were sub charters for my old firm, Bahamas World Airways, from Brussels to Gander. The aeroplane then went on to the islands while I rested in Gander before bringing it back. Alas, old habits die hard, and soon it became clear that, as usual, Bahamas World didn't consider prompt payment of invoices a matter of any importance and that work folded up. British Midland had been offered the same contract but, fortunately, the penny had dropped before they began painting the dolphin logo on the Boeing tail fin.

My final Transasian trip was to ferry an ancient Boeing from Luton to Florida. I was on standby for the job while the aircraft was made ready. Eventually, I got an urgent call asking me to get to the airport as soon as I could. Some hours later we were running the preflight check list and completing the paperwork when an operations guy came running across the ramp with the urgent message to get going as soon as we could because a court bailiff was on his way to slap a writ on the aeroplane. It seems that a few days previously it had suffered a total hydraulic failure on a test flight. The undercarriage had been locked down using the emergency system, but all along the approach the aircraft had been trailing a cloud of hydraulic fluid. The result was about a million pounds worth of damage to vehicles in a car park under the approach. From that moment, everything became a blur, and we were taxiing within minutes. There was a minor panic when it was realised that the buyer's agent had left his briefcase in the passenger cabin, so I had to open the passenger door again and lower his bag on the escape rope. When the bailiff announced himself in the front office, he was calmly told that the Boeing was climbing through 12,000 feet somewhere near Birmingham.

Even then the fun wasn't completely over. We couldn't get the flight level we wanted because the Boeing didn't have inertial or Omega navigation, so we had to plod along almost stubbing our toes at 28,000 feet. The aeroplane was really a wreck, the number four engine wasn't up to the job and had to be run at reduced power. I had read all about Dutch roll in high-speed jets, but this was the first time I had actually experienced it and it was quite frightening. Especially if you know how many bolts hold those engines on!

We hit Gander in a snowstorm and near white-out conditions. It was so bad that a Britannia in front of us got lost and couldn't find their way off the runway. We had to make a go around and start the approach again. We finally made it to Miami that night but some clown in air traffic gave us a wrong instruction, so that the skipper taxied the wrong way into a dead end and, of course, you can't reverse taxi a Boeing 707. To end a perfect day, we then got snarled up in traffic caused by a motorcade President Jimmy Carter's wife was travelling in.

We got into even more trouble coming back, with another Boeing 707 that we had collected from American Airlines' engineering base at Tulsa, Oklahoma. That was an aeroplane built specifically for operation within the United States with no long-range navigation capability at all. Not even a seat or desk for a navigator, let alone a sextant mount in the flightdeck roof. It was needed in a hurry and there

N725CA, our clapped out, dutch rolling Boeing 707 at Gander, Newfoundland in 1978.

were delays in the legal handover process, so there wasn't even time to adjust the two compass systems which didn't agree with each other. There was nothing for it other than to fly the Atlantic following the flight plan headings and using whichever gyrocompass system agreed best with the two emergency standby magnetic compasses, and neither of those had been adjusted either! Once we were beyond range of coastal radio beacons all I could use for position fixing was one of the few remaining wartime CONSOL stations.

When radar picked us up approaching Europe, we were about a hundred and twenty miles south of track. Thankfully, there was no safety risk of collision because, with the aircraft being so light, we were way up above all the commercial traffic. All the same, radar reported us, and the chief pilot received a very angry letter from the authorities. Fortunately, there was no way they could take legal action against us, so all we could do was send a very contrite letter of apology, assuring them it wouldn't happen again.

That pleasant summer rolled by and as autumn became winter, BMA offered me a six-month contract on their newly acquired ex QANTAS fan-engined Boeings. It was an all-cargo contract flying between Amsterdam and New York for Pakistan International Airlines. Three of we navigators covered two weekly round trips between us, which conveniently gave me some longish stops in New York and a day or two back home, where I carried on doing a little light aircraft instructing when the weather was suitable.

In New York I would get the train to Amityville and rent a Piper Cub for nine dollars an hour, building up my tail wheel experience. I could also rent a small Grumman and fly up and down Long Island, which was a big help in adding to my night flying hours and much cheaper than back home in the UK.

The work itself was like a paid holiday because each week I had to operate one way and travel as a first class passenger the other way. One time I flew home first class on Friday and went back the other way first class on Monday. Usually, I was off on a Monday, so worked over at the flying club. Then I'd have supper with my girlfriend, say goodnight to her and drive down to Heathrow overnight, nodding off in a lay-by if I felt like it. I'd leave my car at a nearby hotel we had a deal with, get some breakfast, then join a flight over to Paris in time to catch the afternoon PIA DC-10 flight to New York. By then I was pretty tired, but usually we had a complete row of first class seats to ourselves. I didn't much care for the PIA flight catering, so I'd just ask for a Camembert Cheese, some tins of diet Pepsi and a bottle of Drambuie. Mixed fifty-fifty, that makes what is known as a Riyadh Scrambler, but don't ask me why! Each time I woke, I'd munch a bit of the cheese and wash it down with the Scrambler. Sheer luxury! It was a tough life, but somebody had to do it!

The working sectors just involved setting up the new OMEGA navigation system and monitoring it. Usually checking it against a couple of three-star astro fixes on the way. I didn't even have to do the flight plans—the Pan American computer department did those for us.

As the end of the contract loomed, it became clear I was

unlikely to find any more work as a navigator but then, quite out of the blue, an offer as a C130 Hercules navigator came from the Royal New Zealand Air Force.

Chapter 19 - RNZAF

Although I had about a thousand pilot hours at the time, I'd never really wanted to be a pilot, but it seemed a convenient alternative. With my experience, they were not bothered at all about my right eye, which had not got any worse over the years. I was offered the rank of flight lieutenant with guaranteed promotion to squadron leader after two years. The downside was that I would be grounded at the age of forty-five though guaranteed a ground job for as long as I wished.

It all sounded great, but in practice it didn't work out. First, NZ Immigration wanted evidence of my divorce. I had a photocopy of the decree absolute, but I had binned my copy of the decree nisi. I was not in touch with my ex-wife, so it took some time to get copies from the Crown Court. They were even a bit difficult about copies and not originals, and then they wanted written evidence that I had no children. Notwithstanding the fact that we had only cohabited for one year, that's not an easy thing to prove. In the end they had to be satisfied with a letter from the Clerk of the Court saying that to the best of their knowledge I had no children. This caused a delay of at least two months, so I missed the start date of the annual C130 Hercules course and had to accept a posting to short range Andover aircraft instead.

I was formally sworn in by the NZ Defence Attaché in London and a day or so later I was on my way to Christchurch. Fortunately, I'd been able to get a special price ticket thanks to British Midland Airways, since the RNZAF wouldn't pay my fare! True to form, air force life always seems to be 'Hurry up and wait', so I arrived at Wigram Air Force Base in the dead of night, thanks to a delay in Sydney. Nobody was expecting me, nobody had even heard of me and the only identification I had was a letter from the Defence Attaché in the UK. Anyway, the Orderly Officer found me a room in the mess, and I guess I slept the better part of twenty-four hours. It had been a long drag out via Bahrain, Singapore, Darwin, Sydney and Wellington. In fact, I'd been lucky to get in and out of Australia without being arrested since none of the immigration staff seems to know anything about the law that RNZAF Officers did not need an entry visa.

I spent about six months on a holding posting in the Navigation School, as a sort of unit odd job man. I invigilated examinations, marked papers, sat some exams myself, attended some lectures and exercises round at the Command Training School and did a bit of flying in the de Havilland Devons. Those didn't have full dual control and the right seat was normally occupied by an air signaller. However, there was a control yoke and rudder pedals, so I could always 'pole-snatch' and do the landing. Trouble was, although I could fly the old Devons fine, I was having a hard time with the navigation. From the back seat of a B707 in smooth air at nearly 40,000 feet, with all the latest navigation aids, it was quite a shock to be lurching around at 1,500 feet over the sea with no autopilot and an array of aids which could have been found in a World War Two bomber.

One Friday evening, a group of us were shooting the breeze around the mess bar, when a grinning ground branch flight lieutenant, who was known to generally dislike aircrew, came in waving a teleprinter message saying that due to the critical manning situation it had been decided that no aircrew would be permitted to resign for another two years. Too many pilots had quit to join the airlines. I was unsure if that applied to me or not, and I went to see the Wing Executive Officer first thing on

Monday morning. He got on to Defence HQ in Wellington for guidance and later that morning we had a signal that I must resign right then, or I would be locked in for two years at least. Within a week I was a civilian once more.

I didn't want to waste my air fare home so spent a month seeing the sights of New Zealand while I had the chance, then came home via Los Angeles and drove a rental car to Fort Worth TX. I did a bit more recency multi-engine flying at a friend's flight school there and hopped the famous Braniff Airlines 'Big Orange' Boeing 747 to Gatwick and home.

Another matter that had affected my decision to come back was that UK airlines were suddenly short of pilots and doing a lot of hiring. Also, Mrs Thatcher winning the election made the future look rather rosier than it had in the Wilson and Callaghan years. With the exemptions, thanks to my American pilot qualifications and my UK navigator licence, I had very few tests to get my UK CPL/IR issued. Unfortunately, by the time I got those out of the way, a couple of airlines had gone broke, and the market had flipped. In the end, the best job I could get was as an air taxi pilot flying Piper Aztecs.

Chapter 20 - Air Taxi Pilot

I f I'd thought that some of the old-time charter airlines were rogues, I had something to learn when I got involved with an air taxi company. The boss wanted me to fly VMC all the time, and even cancelled IFR flight plans after I had filed them. The Chief Pilot, a former RAF Lightning pilot, had been 'removed' as 'unsuitable' by the CAA and was operating as a training captain. On one occasion, he briefed me on what to do if an engine quit on a take-off above the maximum permitted take-off weight. I told him I wouldn't have taken off overweight in the first place, which didn't go over well at all. Fortunately, the new Chief Pilot, unlike our Lightning ace, was an experienced airline pilot, and I could always rely upon his support.

Small air taxi aeroplanes are not required to use licensed aerodromes, so some of the places we went into were somewhat hazardous, especially in poor weather. I guess it was a bit like bush flying in third world countries. One place we frequented could mean following a meandering river under a low cloud base and doing a steep turn ninety degrees left when you saw a certain wood. If the 'runway' wasn't in sight, it was pour on the coals and climb fast as you can because there was a 2,000 foot hill just a few miles away. Half the aerodrome had been scooped away as a gravel pit anyway, so there wasn't a lot of the runway left.

Our engineers were a great bunch but were never allowed to clear all the snags and keep the aircraft in really tip top condition. I was lucky in that I never had a major problem, although a friend did have an engine blow up on take-off. That was so bad that shrapnel punched holes in the cowling on its way out. Happily, the engine must have seized so the prop feathered itself somehow. The firm would charge the customer for the use of a handling agent at some airports, then avoid using one. What we had to do in a one hour turn round at some places was quite impossible to achieve, especially when parked some distance from flight briefing and the Met Office. On top of all the other problems, a few of the passengers were quite objectionable and rude, expecting us to carry their bags and so on. I assume they were accustomed to being such big noises in their companies they treated their own staff the same way.

I think single crew air taxi is the toughest job in civil aviation, apart from the North Sea helicopter boys. If you can do that you can handle any job. I wasn't much good at it and fell out with the boss and the chief training captain once too often. Maybe getting fired was the best thing that could have happened, except that it condemned me to another three years out of the airline business. A few months later, I was sitting in a small Cessna a few thousand feet above our airfield, when I heard our ex fighter ace training captain call on the approach frequency. I could see down through the mist but, of course, it was not so easy for him to see through it. Eventually, I spotted his large twin Cessna lined up on a disused runway about two miles away and very close to some high ground. I hit the transmit button and called 'You're on the wrong runway Alan. The airport is two miles north of you'. Seconds later I could see that he was climbing away and then came a hesitant, 'Ah, yes. I see. Thanks for that.' came over the radio. About four years later he flew into a hill in IMC and killed himself. Mercifully, with nobody else on board. It was a classic mistake: he forgot to reset the altimeter. QFE and QNH again— remember? I still feel sad about that. He may have been a cowboy, but he really was quite a nice man.

Fortunately, I was already planning to leave. The owner of another airfield where I was still doing some instruction had bought a Piper Chieftain and wanted to set up his own

Air Taxi captain at Aberdeen on a windy day in 1980.

air taxi operation. I had been quietly working away on the paperwork for the Air Operator Certificate application. The CFI could fly it himself because he had no instrument rating, so I had been offered the job. Alas, the big boss, a very wealthy Midlands industrialist, was rather headstrong but not really a very good pilot himself. Without telling anyone, he took the Chieftain to the South of France for the weekend. The first I knew of the disaster was the black looks all around when I pitched up on the Monday morning. The Chieftain had an engine quit on the climb out from Nice and crashed killing the boss and his wife. Many times we had told him not to lean the mixtures off quite so much, but he kept on doing it and blew a cylinder off.

I ended up doing any flying job I could. Flying photographers around at low level, ferrying aircraft, freelance air taxi work, instructing, and now and then the odd job as second pilot in a corporate Beech King Air.

Chapter 21 - Instructor Again

The airline hiring market was very flat, but the owner of one of my old Britannias came up with an offer, of sorts. If I would self-study the type rating and pass the technical exams on the Britannia, his new company would give me the flight check and put the type on my licence. I think the Britannia technical course had been about six months full time in BOAC. I borrowed a set of notes, memorised a lot of questions and answers and passed it in six weeks. I believe I was the last person ever to take the test for there were few Britannias around by then. That, however, was the good news.

The plan was that their sole aeroplane was scheduled to arrive at Bournemouth and unload a cargo from West Africa. The next day it was due to be ferried to Manston, Kent. I was to sit in the right seat, suck up the gear and the flaps and do the radio, etc. Then I'd be signed off as a competent copilot and fly as a third pilot for a few weeks until I really did know something about the job. It all sounds a bit casual, but that kind of thing was common in those days and, somehow, it worked well enough. Anyway, the bad news is that the ferry to Manston never happened. The firm had already gone broke.

I did one last flight as a navigator in 1983. Heavylift Airlines approached me with a view to flying their CL44 'Guppy', a much-modified aeroplane for carrying outsize loads. An Irish airline, Aer Turas, allowed me to fly as second navigator to New York and back, to revalidate my licence but, in the end, Heavylift did not get the contract they had hoped for. That was soon after the end of the Falklands War and they had been hoping for a contract to fly a lot of supplies out via Ascension to Port Stanley.

Some amusing incidents took place at one location where I was instructing. A certain private pilot, not the sharpest pencil in the box, was trying to clock up enough hours to get a commercial licence. He'd failed the instructor entry exam, so was trying to get seven hundred flying hours any way he could. I'd been down at the maintenance hangar one day and saw the mechanics fitting a huge ferry tank into a Cherokee Six. That's not unusual. I guessed it was being readied for a North Atlantic delivery ferry flight and thought no more of it.

Meanwhile, our friend was approached by someone and asked if he could do the flight plans for a trip to Morocco and back. A suitable fee was agreed, and he duly set to work. When he delivered the paperwork, he was asked if he would care to make the flight. With visions of being paid for about another twenty hours in his logbook, he agreed. But when the other chap said, 'Do you realise what you'll be bringing back?' he realised he'd be getting deep into drug smuggling. He refused, took his fee and left, having been given a warning not to say to anyone what had gone between them.

A few days later, I heard reports of an incident on a disused aerodrome about twenty miles away. The aeroplane with the huge ferry tank had landed there but, as it rolled to a halt, a squad of heavily armed police emerged from the bushes. The pilot did a quick about turn, gunned the engine and took off again, whereupon a Police helicopter made a sudden appearance. In those days, just about all professional chopper pilots were ex-military, so the light aircraft pilot didn't have any chance of outwitting the chopper pilot. He was eventually forced down in a field twenty miles away, with the helicopter crewman videoing the small packages being thrown out of the direct vision window. The pilot was arrested, along with half a dozen other individuals who hung around the airfield, including our flight planning expert. I believe the case against him was eventually dropped, but not until his family had taken out a second mortgage to bail him out to the tune of £20,000!

After a few false starts I did land a job as CFI of a small flying club, which at least paid the bills for a couple of years. In fact, I quite enjoyed a lot of the flying and made a lot of friends. The only drawback was that the owner fancied himself as a high-flying management consultant and tried to apply some of his bright ideas to the club. Ideas like lashing out nearly £600 for a fancy cash register, when all we needed was a simple invoice block. Then he hired a 'Sales Manager' who was a complete clown and knew absolutely nothing about aviation. Inevitably, we got badly behind with fuel payments, so the airport manager had us closed down. Not that I was too bothered because I was just about to put my notice in—I'd been offered a job in the CAA. When it got back to the airport manager that I was joining the CAA as an Airport Inspector, I was reliably informed he almost had a cardiac arrest.

Chapter 22 - CAA Airport Inspector

The CAA job was quite fascinating and although it didn't involve much flying, at least the pay was better than I had got used to in general aviation. Pilots take aerodromes for granted, so there was a lot to learn about the subject. I was based in central London which wasn't a problem for I had the use of a friend's spare room a few stops along the Piccadilly Line and went home every weekend.

About one day a week, on average, I was out doing some sort of inspection at Heathrow, Gatwick, Luton or some other airport in Southern England. Now and then I could get a jump seat ride on the navigation aid calibration aircraft based at Stansted and the Authority guaranteed to fund enough light aircraft flying to keep all my licences current in case they did need to post me to a full-time flying job. In practice, I was able to do enough weekend instructing to keep my commercial licence valid, so it was agreed that I could spend the cash allocation on whatever flying I wished. I blew it all on getting type ratings on the Beechcraft King Air and the Cessna Golden Eagle. It seemed a good enough idea at the time but, in fact, I never flew either type after I passed the tests. There was even enough left over to rent an Auster for sufficient hours to get a chum up to solo standard.

The part time instructing nearly got me killed on at least one occasion. I was being checked out to instruct in a Tiger

Moth, in which the instructor sits in the front seat, as the lower wing prevents a good downward view. I had called downwind and was cleared to final approach. I had turned on to final and was cleared to land. At about the height of a double decker bus I had begun the flare and reduced the power, when the chap in the back seat suddenly grabbed the stick, shovelled on full power, and started to climb. I never saw the Piper Cub that had somehow flown down the visual approach directly below us.

When we got down, I went to find him and speak to the duty ATC controller. It seems the youngster in the Cub had launched from a nearby RAF Station for an hour of local flying but couldn't get back because a thunderstorm was sitting over the place. He diverted to our airfield, even though it was listed as not available to non-radio aircraft. In a case like that, he should have made an 'overhead join' to the circuit, which would have enabled ATC, or us, to spot him. Unfortunately, he was he was low on fuel and didn't think he enough fuel left unless he made a right-hand join onto final approach. Pretty bad airmanship and completely unlawful anyway.

What he should have done was make a precautionary landing in a large enough field, of which there were plenty nearby, while he still had enough fuel to do it safely. After all, precautionary landings with power are part of basic flying training. I think he had 'learned his lesson' though, and although we had to file an 'Airmiss' report, my friends in the Air Safety Enforcement section decided not to take any action against him. It would have helped if the controller had been keeping his eyes open too. He'd had a perfect view to the left and to the right of the approach track, yet he had cleared us to land having never noticed the two aircraft so close to each other.

Another close call involved a Tiger Moth I was ferrying for a dealer who was trying to sell it. I was on my own at about 6,000 feet on a warm summer evening, perfect for a few mild aerobatics. The aircraft had previously been rolled on its back and three wings had been rebuilt. I was lining up for a loop along a disused runway below me, just diving a little to build up speed, when I remembered the aeroplane wasn't cleared for intentional spins. Being cautious, I just levelled off and carried on. When the aircraft was sold, the new owner had the other wing opened. They found the main spar was cracked in three places and one of the cracks had almost gone right through. It still gives me the shivers, especially as I wasn't wearing a parachute.

That reminds me of a big Cessna single I used to fly dropping parachutists. A colleague recovered from a spin at about five hundred feet with five parachutists still holding on for dear life in the back. Centrifugal force kept them away from the exit. There was always a danger of inadvertent spins because, at each drop, the aircraft had to be flown in flat turns at close to stall speed. Later, that aeroplane was lucky not to kill anyone. I was asked to take over after one lunchtime but was told there was 'slight damage' to the tailplane. On examination I found the tip of the right side was bashed about four inches out of place and the spar was bent. Even the elevator wasn't working as it should, yet someone had been flying it all morning and the owner even flew it away to have it repaired. Someone must have done a very tight and fast turn when parking it the night before. There was clear evidence that the tail had smacked a concrete fence post good and hard.

After about a year I got wind that my next CAA posting would probably be to a section I wasn't very keen to join. It would have meant promotion, but it would also have meant staying that backwater until I retired. I was also getting tired of commuting into Central London and spending most of my time in digs. At that moment, however, the chance of an airline job finally came up. The start date was a few months ahead, so I temporarily joined Cabair Limited, to set up a new flying school at Blackbushe Airport.

Chapter 23 - Airline Pilot at Last

I pitched up on day one to find an empty office with a telephone on the floor. Within three weeks I had a three-aeroplane flying school up and running and two full time instructors working for me. I stayed for a few more weeks, then handed over 'command' to one of the assistants I had hired. Unlike so many flying schools and clubs I had flown for over the years, Cabair was a very slick and professional organisation. It had originally been the London School of Flying, where I had worked back in 1965, but had flourished under new ownership becoming, at one stage, the biggest light aircraft operator in Europe. It was spread over three bases at Elstree, Denham and Biggin Hill, and it encompassed helicopter, air taxi, aircraft sales, and engineering divisions.

We leased part of the old terminal building and started off with three four-seat Grumman Tiger trainers and one Slingsby Firefly aerobatic trainer with two seats. Company ethos was to provide a quality product with no short cuts or cheapskate operation. Aeroplanes were always clean and tidy, always parked in a neat line. Staff wore neat uniforms and emphasis was placed on proper ground school training before every flight. Cabair's services were not cheap, and they were never intended to be. The aim was to attract the type of customer who would typically be a successful professional in early middle age, with a good disposable income but who's spare time was valuable to him or her.

At the end of the course, it was hoped we would be able to sell them an aeroplane. If they bought it, they could lease it back to us and it would get good utilisation. That way, both the student and Cabair would benefit. In other words, the attitude was 'If you can't afford to fly with us, then please go somewhere else.' Harsh, maybe, but it worked long enough to keep the firm in profit for forty years. Sadly, Cabair faded away just a few years ago, albeit 'under new management'.

I finished at five on a Sunday evening and started a Herald ground school course at nine on Monday morning. That lasted two weeks, then I was packed off to Birmingham to revalidate as a Piper Navajo captain while we awaited my exam results from the CAA. This included, by the way, passing the AC electrics exam for the third time, because passes only lasted six months unless you passed the flying test as well. I still don't really know anything much about AC electrics and I still can't figure out how I passed when a friend with a full PMG Radio Operator certificate failed.

The plan was to employ me as a Herald First Officer as well as a Navajo captain. I passed all the base and line checks on the 'Jo' yet never flew it again for them, being too busy flying the Herald. Meanwhile, the test results for the Herald came through and I began training on that lovely old truck. Life slowly settled down to flying overnight cargo schedules for Securicor between Birmingham, Brussels, Belfast and later Dublin. For most of the next twenty years I flew almost entirely at night. One year I didn't fly in daylight between August and April, but I was still able to do a bit of dual instruction in light aeroplanes. With a bunch of friends, I started a flying club at Wellesbourne Mountford, my old RAF Station, and it's still in business now.

The work was a rigid routine, and the aeroplane was never available for charters. Maintenance was high quality and we encountered very few delays and cancellations. The entire operation was very slick and professional and all the captains I flew with were highly experienced operators one could really learn from.

One amusing incident involved four million pounds in used notes. We were hauling those back for destruction in London. As usual, we called on the company frequency as we passed Wallasey, giving the estimated time of arrival. We were pretty slick at arriving on the chocks within a minute of ETA, and we rolled to a halt right on time. As the engines wound down and the door opened, the warehouse foreman entered with news that the armoured truck hadn't arrived, and we would have to remain on board to guard the cargo.

My skipper made it clear we were not paid to be security guards and they could darned well look after it themselves. Consequently, four million quid in used notes spent half an hour stashed in the office broom cupboard. The following week it happened again, right under the noses of the main board directors, who happened to be on a visit to see their aeroplane. They soon put that right and the armoured truck never came late again.

Another time we were checking the cabin before an empty ferry flight when we found a small package of industrial diamonds hidden in a cargo net on the floor. Probably a few million more pounds just lying around. It wasn't worth heading for Cost Rica, so we turned it in.

The 'Night of the Great Storm' in 1987 was something not easy to forget. Thankfully, the wind was down the runway at base, destination, and the various diversion fields. The old Herald usually tanked along at around 170 Knots, but that night we were showing a groundspeed of just over 300 Knots. Great, until we headed home, and at one time it was down to 78 Knots. There was a similar storm a couple of years later and the into wind runway at Cologne was unavailable. I could almost hear the copilot praying 'Please God, make him divert' but, in fact, I didn't really need any divine pressure. We diverted to Düsseldorf where the runway was right into the wind, but it took a very long time to cover the fifty miles or so. Even then, our problems were not over. It was gusting seventy knots, which is twenty knots more than you could legally even taxi a Herald. We got it parked into the wind, somehow, but it was a real fight fitting the external rudder lock.

Once or twice, I was tempted to leave and progress to a better type, but the boss always persuaded me to stay. After a couple of years, I was made up to captain. Sadly, six months later, Securicor decided to close us down. They were very pleased with our on-time departure record, but had simply decided they didn't want to be in the airline business. They owned our one aeroplane so, alas, we had to go. We were replaced by two companies, in turn, who made a complete balls of the contract and both of them went bankrupt.

The Securicor flagship Dart Herald G-AYMG approaching runway 33 at Birmingham as I renewed my instrument rating. Photographed by a friend using my camera.

Chapter 24 - New Airline

Happily, it was one of those all too rare periods of severe airline pilot shortage, so we were head hunted by the airline that bought the aeroplane. They were so keen to get the six of us they sent their chief pilot over to Birmingham to meet us, rather than ask us to visit their base for interviews. We had to do some training at Bournemouth, though we continued to be based at Birmingham and flew mostly to Cologne on a UPS contract, as well as Brussels for Federal Express. At roughly six-week intervals we were rotated, for a week, to another base such as Luton or Bournemouth. Everyone got a bit of variety, including the crews at other bases.

Although the aircraft maintenance and crew scheduling were not quite as good as at Skyguard/Securicor, generally the firm was very well run, and the pay was much better. The worst thing about it was that the duty hours tended to be much longer, days off in short supply, and we were doing less flying. The first black I put up was coming back off holiday with a broken leg. Nothing so exotic as skiing—I just tripped on an uneven pavement in Canada. That put me off sick for seven weeks, but they were very generous about it and continued to pay me.

The year before, I had bought myself a small aeroplane. A Luscombe Silvaire, 'The Sports Car of the Air', as it had once been advertised. That seemed a great idea at the time, but with the shortage of rest days I was rarely able to fly it. I owned it for eight years but only managed about thirty hours a year. It was great fun, when I could fly it, and I was able to visit airfields I'd never been to before. I liked to hop rides with it, and over the years took nearly two hundred people from about age five to ninety-five up in her. Because I was able to keep it registered in the USA, it was very cheap flying, as I was able to do almost all my own maintenance. In exchange for odd jobs that I did for the airport manager, I never had to pay any for any parking or hangar accommodation. Major expense, apart from fuel and oil, was insurance, but that turned out to be good value for money when I eventually wrote her off. The left brake suddenly locked on at about 30mph at the end of a landing roll. With the aeroplane heading towards a barbed wire fence there was nothing I could do but give her a 'boot full' of right brake, which ground looped her. I'd have got away with that, if there not been a grass hummock at the edge of the tarmac runway which she skidded sideways into, breaking the undercarriage leg, shock loading the engine and breaking the wing spar. The insurers paid up the agreed value which was, fortunately for me, more than I'd paid for her eight years earlier. Happily, her remains were acquired by a friend, and she will be flying again soon.

There were a couple of times when we got into Southend, or diverted to Stansted early in the morning and the customs took so long clearing the cargo that the operations manager wanted me to extend the duty period and continue to Birmingham. Not having had a proper night's sleep for nearly a week, I refused to extend our duty period because it was clearly a safety issue. Both the first officer and me were too tired and fatigued and even if the numbers had added up to a legal flight, it would have been an offence to carry on. The Chief Pilot had no choice but to back me up because the company manual clearly stated that captain's discretion to extend a duty period must not be used to 'achieve an unrealistic commercial schedule.' Quite remarkable that, within days, a manual amendment was issued removing that statement. I don't think our CAA flight operations Inspectors should have permitted that.

I suppose I had made a problem for myself by staying thirty miles from the airport, although getting there after the evening traffic was not difficult. The worst problem was driving home tired in heavy morning traffic, and I narrowly missed a couple of serious accidents. It was worse when we were asked to leave the aircraft at Southend and get ourselves back to base in a rental car. Some pilots flatly refused to do that, but I usually agreed because the cab drivers often seemed to be even more exhausted than we were.

There was quite a serious problem with the propeller feathering motor on the Herald. If you couldn't cure it, it meant shutting the engine down. Any twin-engined airliner on one engine is an aeroplane in serious trouble, and it just had to happen ten minutes after take-off at maximum weight from Birmingham with the weather just above take-off minimums and well below landing limits. Mercifully, Manchester was wide open, and about ten minutes ahead of us. The gear and the flaps went down as late as I dared, but the good engine was using a lot of water-methanol injection for the final few minutes. Some 'wise guy' in the company said I ought to have removed a 275-amp fuse which would have stopped the runway motor and allowed us to carry on. He had forgotten it would have made it impossible to feather the propeller if it went into fine pitch which, in turn, would have made it impossible control the aeroplane and we would certainly have died. A training captain I knew tried it in a simulator for a very similar aircraft type, the Fokker Friendship. Four times out of five it 'crashed'.

A double diversion is a situation that any airline pilot dreads. We were headed to Birmingham, which was forecast open, with acceptable weather at our diversion field, East Midlands. Birmingham socked in when we were about twenty minutes out, so we immediately set course for East Midlands, which also began to deteriorate. I made two approaches and had to go around but Birmingham was still out. A third attempt would have been unlawful, so off we sped to Luton. I got into Luton at the second attempt with barely enough fuel for a third try.

One night I was asked to do a flight at very short notice.

I agreed because I was promised I would be taken off the flight the next night. I got to the airport in time and did the flight but didn't sleep much the next day because I knew I had the night off. At ten o'clock, I got a phone call asking why I wasn't at the airport. The message had not been passed on. Clearly, I wasn't in a fit state to even drive to the airport, so I flatly refused. Another black mark. After about two years, I think I was suffering from cumulative fatigue, and when I took a week's leave I spent it almost entirely in bed with bronchitis.

Then I had a couple more in-flight emergencies. We were tasked to fly into Maastricht, an airport in a very noise sensitive area. The chief pilot had figured that if we made a visual approach at almost twice the usual height with the gear down, full flap, and the power back to the minimum acceptable, we could just make it without getting a noise violation. Although that's no way to operate an airliner I did thirty or so of those until they sent along a training captain to 'show us how it was done'. That was the first night we ever got a noise violation. A week later when I was taking off I called for the gear to be sucked up but nothing happened—the gear was firmly stuck down. That slowed the rate of climb, but we had lots of fuel. As the weather was good, I decided to press on to Brussels rather than risk a noise violation going back into Maastricht. Continuing to Brussels would also make it easier for our engineers to sort the problem out. As we approached, it was obvious that I had enough fuel for Ostend, so I re-cleared for Ostend and landed there.

Another Herald, on an empty ferry from Scotland to Bournemouth, was diverted to Ostend and our load transferred to that aircraft. With the gear locked down, I ferried the aeroplane to our main maintenance base. Two weeks later, I was called into the office and almost fired, until I pointed out that the problem was that the gear had failed to go up, rather than come down. In that case the procedure was to continue to destination so long as the weather was good enough to eliminate any chance of a need to overshoot, then select gear down using the

Mayhem -Ostend 1992

Air Marshal Sir Gerry Mayhew, as a young fighter pilot, getting a taste of airline flying.

emergency system on final approach. It transpired that the new Chief Pilot hadn't even read my report; he came up with some lame excuse about being away on a course at the time. It turned out that a sequence valve casting had cracked causing fluid to leak. Something that had never before happened to a Herald.

The final straw came leaving Brum one night. I had snagged the aeroplane for a propeller control fault, but it had been cleared by some minor adjustment. Then, a senior training captain snagged it again for the same thing with the same result. The next time I flew her it happened again, so I put it in the book again. Maybe we would have gone on for weeks like that, but fate intervened.

In Skyguard we made it *de riguer* never to remove the control locks until we were lined up on the runway. However, the previous Chief Pilot, when giving me a check flight, had said it was quite all right to do that. Normally, that would have been okay but, combined with this insidious propeller control snag, it meant disaster. Two aircraft were ahead of us and the first one got airborne okay. The second one lined up and air traffic asked if we were ready for an immediate take off because there was a jet just a few miles out. Naturally, I accepted, and as soon as the aeroplane in front left the deck, we were cleared for take-off. When my first officer advanced the power in the normal way, there was a loud bang, and all the lights went out. At first, I thought a bomb had gone off in the hold, but I could see flames licking around outside and the left propeller had stopped, followed by the other. By then, the emergency lighting had come on and all the bells and whistles were roaring away. My copilot, quite sensibly, abandoned his seat and headed for the door. After all, he was a married man with three kids to look after. I didn't want to evacuate without making sure the aircraft was as safe as possible, so I shoved him back into his seat and said 'let's do a few emergency shut down checks'. I called ATC, although by now they could see something was amiss and the fire trucks were on their way. We switched off everything we could think of and climbed out to find a secondary fire in the right engine. In time, the aeroplane was dragged away, although nobody could find the correct tow bar, so we blocked the runway for long enough to cause some other aircraft to divert. Both engines were ruined at a cost of about £300,000.

I phoned the Chief Pilot, who was just about to go to bed. I told him I was grounded until further notice, which he could hardly disagree with. The first reaction of certain senior staff was just 'Fire the pair of them', but wiser counsels prevailed, and a proper court of inquiry was convened. Perhaps that was necessary before any insurance claim could be filed. Fortunately, the Flight Operations Manager was a former RAF Squadron Leader who knew how to do such things in the correct manner. As his board member, he also had one of our older copilots who had been a commissioned engineering officer in the RAF. The matter was handled impeccably, even taking advice from Rolls-Royce, and the finding was that we were not to blame. The cause of the incident was the fault that we had snagged, combined with the control locks being removed too soon. Because we wore noise cancelling headsets, we could not have been expected to hear the change in engine note well enough. In fact, once again, they learned something about the Herald that nobody had known before. A certain warning light did not come on at exactly the time everyone had thought it did.

I was told I must do a renewal base check with another training captain who had not been involved in the inquiry and then I would be sent off to fly from Paris for three weeks. That was fine until a certain other individual intervened. I was hauled before the Chief Pilot and told that although I could not have prevented the accident I was at fault because I had not used the check list to secure the aircraft before I left and, therefore, he was demoting me to copilot for six months. I couldn't believe this. He was saying I should have calmly sat in an aeroplane that was on fire and used a checklist to switch off things like boost pumps and high-pressure fuel cocks, despite the fact that we had actually switched them all off without a checklist.

Naturally I could not accept this—to have done so would have made me appear to accept responsibility for the accident in the eyes of everyone else in the firm. I was formally terminated and, with union help, filed for wrongful dismissal (it may have been unlawful or constructive dismissal, which are not quite the same thing). The company settled out of Court.

In one firm, I had a real character of a copilot for a couple of years. An above average operator, and great fun to fly with, he shared a rented flat with one of my fellow skippers whose wife owned a tiny dog. One night, our Nigel had been night clubbing with a lovely grey-eyed young blonde. They returned, slightly the worse for booze at about two in the morning and were just getting down to the serious stuff when they both detected a dreadful smell. On went the bedside light, to reveal that doggy had left a little visiting card inside the bed. The pair of them were liberally smeared with dog shit. My word, how romantic can life get!

Chapter 25 - A False Start

An old mate contacted me about an airline startup he was involved with in Greece. Would I care to join as a Boeing 737 first officer, on condition that I would also run the company navigation office? I was promised a command after the first six months. So off we went and did the ground school course which was excellent although I dipped the CAA exam by about two percent. The reason was penalty marking, you lost half a mark for every answer you got wrong on the multi-choice test. That was to discourage you from guessing answers. Stupidly, I guessed a couple of answers, but it made no difference anyway. The financial backers got the wind up over the wars in the former Yugoslavia and pulled out. The airline never did get started.

Fortunately, I was snapped up by another Herald operator and that job lasted almost a year before they laid some of us off. The company hadn't been paying the owner for the lease of the two aircraft, so he repossessed them. The company were not especially bothered because they had found some cheaper Hawker Siddeley 748 aircraft to replace the Heralds. Loyalty is usually a one-way street in the airline business, so we were dumped in favour of some Dan-Air crews who were 748-rated but laid off when British airways swallowed Dan-Air. Airline management so often take the short-term view. Most of those 748 qualified guys also had jets on their licences, so they were gone in months or even weeks and the firm was soon back to training new people on the type.

Notwithstanding some of the maintenance problems we had, I quite enjoyed that flying. Most of it was heading out to Ostend in the early evening and waiting in the restaurant until the trucks arrived. We then loaded up with car body panels destined for the Honda plant near Swindon, and were usually home by two.

Chapter 26 - Final
Approach

I kept in touch with the owner of the two Heralds and, one day soon afterwards, he called to say he had a new customer. Could I get half a dozen crews together? The negotiations dragged on for months, but we eventually we got them flying later that year. They were based at Stansted and flew an overnight Post Office crossover service to Belfast and Gatwick. It went on for about a year and was mostly very pleasant apart from the dump that passed as a hotel near Belfast. One freezing night, I got to my room to find just one blanket on the bed. It took the night porter half an hour to find another, with me threatening to call a taxi and ship out to another place. Although the food seemed okay the place was cold, dirty, and noisy. I kept up the pressure to move us to the airport hotel which would have cost no more as we would not need the taxis. Unfortunately, I was constantly undermined by one of the copilots who had designs on the body of the receptionist—a young lady well known to be 'available'. He kept telling the bosses back at base what a fine place it was. In the end, the Chief Pilot, who was on the other fleet, came over to see what it was all about, and he was appalled. Right away we moved to the decent hotel.

Disaster struck one freezing night, though mercifully not at us. We launched from Stansted and were a few minutes from Daventry beacon when we went into cloud at 12,000 feet. The de-icing system wasn't making much impression,

195

so I asked for a descent to 10,000. At ten we were still collecting ice at a fair old rate so, after a few minutes, I asked for 8,000. Approaching Wallasey in desperation, and probably starting to look a lot like an airborne iceberg in desperation, I asked for 6,000, although it was below the airway base level. The controller assented but commented 'I assume you have read the Sigmet?' (a Sigmet is a signal to warn pilots of very severe weather conditions). I twigged and replied that I had done so, though no Sigmet had been given to us at briefing.

Meanwhile, unknown to us, a Viscount aircraft flying south from Scotland to Coventry had crashed in Staffordshire killing its Australian captain and severely injuring First Officer Ray Coles who, years before, had been one of my assistant instructors at Blackbushe. It seems that they hadn't been given the Sigmet either, or perhaps no Sigmet had been issued.

Aviation is often slow to learn from its mistakes. That was the very same cause of the loss of the Redcoat Air Cargo Britannia near Boston, Massachusetts in 1982, killing two old mates of mine (I'd originally been offered the chief navigator job at Redcoat, but turned it down). At the time of the accident, they only employed two— so I'd have had a fifty percent chance of being on board. The next time I went into the office at Stansted there was a file in an obvious spot on the counter with SIGMETS in red letters, as large as you could write them At least somebody was trying.

I visited Ray in Newcastle under Lyme Hospital soon as I found where he was. He was in bad shape and couldn't see me, although he knew I was there, and we could talk. There wasn't anything I could do to help, but his airline was very good to him. The hospital patched him up and within a few months he was flying again. I've not heard from him since, but I believe he is still flying somewhere as a jet captain. Good going for a guy who, like me and so many others, started out as hangar mechanics. The real cause of that disaster was not just the icing. Another reason was an emergency check list badly printed on pink paper, and almost impossible to read in poor lighting—so nobody had ever bothered to use the checklist. In fact, Ray was even luckier than he thought. As it crashed, the Viscount had just cleared some high-tension cables. Hitting those he would have had no chance at all.

One of the last Herald flights I did was a very close shave. We were the final aircraft to leave Stansted before the runway closed for overnight maintenance work and the place was, anyway, fogged in. I got a fire warning on the number one engine as we climbed away. Turning back was not an option. We were heading for Gatwick, which was still open, but the closest diversion field was an hour away at Manchester. I called for the emergency check list but as I reduced power on that engine, the lights went out and the warning bells stopped, so I suspected a false warning of course. I levelled off and had the copilot make a 'Mayday' call, telling ATC we had a really bad problem. They gave us a direct steer for Gatwick and cleared us as number one in traffic. I opened up the taps on the good, number two, engine but, if I increased power on number one, the warnings came on again, so I had to baby that one as best I could. I shut it down immediately we landed and taxied to the ramp on one, escorted by an impressive fleet of fire trucks. Meanwhile, another Herald landed a couple of minutes after us and suffered a complete hydraulic failure, blocking that single runway for twenty minutes. The engineers found the fire had burnt through the heat shield and could have got a lot worse if I had kept the engine running for much longer. Maybe my reaction wasn't the correct way of handling that emergency, but it worked, and I certainly didn't favour the idea of diverting all the way to Manchester on one overworked engine.

About that time our company made a bid to renew the contract for another two years, using Fokker Friendship equipment. Plans to buy two late model aeroplanes in very good condition were almost complete, so I was given a pile of study notes and told to go and take the technical exams on type. I passed the test without a problem, but then the CAA stepped in with a list of requirements before they would accept the two aircraft onto the UK civil aircraft register.

To comply would have cost about £70,000 per aircraft, so the company tried to negotiate. Unfortunately, while that was going on, another buyer snapped up the two Fokkers so we never got the contract. With no work in prospect for the two Heralds, it looked like I was going to be out of a job yet again. Then the company came up with the offer of flying a Shorts 360 on an overnight schedule from Exeter to Belfast for the Royal Mail. The job market was pretty bad, and nobody needed Herald rated pilots anymore, since there were only a handful still flying. Our two never flew again and were scrapped, although they were good for many more flying hours.

It was Hobson's choice, and I was given a pile of notes and manuals on the Short 330/360. With that on my licence, I moved to the Southwest for the last few years of my so-called career. A friend had just taken the lease of a pub half an hour from the airport, so that took care of the accommodation problem for the first two years.

I can't say I ever enjoyed flying the Shorts 330. It was replaced by the larger SD360 after a few months, although that wasn't much better. At least the Shorts were good economic performers and the engines quite reliable. We still had our anxious moments though. My old nemesis, an over speeding propeller, returned on one occasion. Thankfully, it happened at low power about two miles out on a visual approach. All I could think of at the time was just do nothing—don't touch the prop controls at all. That turned out to be the right course of action and I managed to get the thing back on the deck in a fairly dignified manner. I don't remember if there was anything in the check lists or the manuals about overspeeding propellers, but I don't think there was.

The GPWS (Ground Proximity Warning System) was introduced in the seventies but by the nineties it seemed to be giving us an awful lot of spurious warnings. I took off in broad daylight one evening and the voice started yelling 'Gear, too low. Gear, too low …' I was already climbing on full power with the gear up. What did the darned thing want me to do? Put the gear down again? I should add that we were climbing over flat terrain with no obstacle in sight.

Another time I was lined up with the Instrument Landing System (ILS) and I was level at exactly 2,000 feet, as required by the published procedure. It started shouting 'Pull up, pull up...' so I had to waste ten minutes flying time starting the procedure all over again because we were in cloud. That was the correct procedure. It's all too easy to sit there trying to figure out why you think it's giving a spurious warning when it's right and you fly into a hill.

Another time about fifteen miles out on the ILS localiser, I was level and above safe altitude, but it began blaring away. On went the power and we started a go-around. But after a few seconds we came out of cloud, and I could see the runway. We intercepted the glideslope from below, followed the ILS under visual conditions and made a normal landing. At first, I thought my boss was going to chew me out for not making a complete go-around, but he saw the sense in my reasoning that if I'd done that, we would have probably got the same spurious warning all over again. He was a good boss—another 'retread navigator' like myself.

Some copilots had minimal experience and I sometimes felt I was still teaching them to fly. One guy was hand flying in some quite bad turbulence one night. Instead of flying attitude and riding the turbulence out he was trying to maintain an accurate altitude and starting to lose control altogether. I took over and got things under control, but he was worried about an 'altitude bust', i.e., getting more than 200 feet from our assigned level and risking a collision. A deviation of more than 200 feet would send off bells in the air traffic control centre. Surely common-sense dictates that maintaining control is more vital that worrying about an altitude bust, especially since we were probably the only aeroplane over Wales at two o'clock in the morning. Besides, if the warning bells went off in ATC, it was my problem to deal with it, not his.

Another copilot was a complete and utterly messianic

'aviation nut', the type who probably got into bed on his wedding night, produced an aircraft technical manual and began trying to impress his new wife with statistics about the aeroplane. He was a great guy though, who had fought off enormous medical problems to get his qualifications. I reckoned he was an above average aircraft handler, and his technical knowledge was encyclopaedic. On a rainy night, when the clouds were down in the weeds with a thirty-knot crosswind, I preferred him in the right seat above any of the others. Sadly, he was no politician and kept upsetting everyone because he couldn't keep his mouth shut. As the saying goes, 'The one thing an Englishman can't forgive you for is being right.'

Modern colour radars are not the complete answer to turbulence. Despite no trace of anything untoward on the screen, one night we punched straight into a storm and got thrown around all over the place. Eventually, something set off the stall warning and the stick shaker. That's a device that makes the control column vibrate rapidly if you are approaching stall speed. When we got back to calm air it was still shaking, so all I could do was pull a circuit breaker to stop it. After we landed, I pushed the CB in again and we got the stick shaker vibration again as we taxied to the ramp.

With the Shorts I never liked to do gear down ferries because the manual said the ground locks should be left in place. Personally, I would have preferred to do it with the locks out so, if I lost an engine, at least I could reduce drag by sucking the gear up. Better to land gear up on an aerodrome that have to put it into a field with the gear down. So much flying safety is basically common sense.

Manufacturers are always telling us how reliable modern engines are, but one other captain almost had a double engine failure on initial climb. There was an ice build-up in the intake that he could not have seen from the ground. As soon as they got airborne the ice was blown back into the engines. There must have been some tense moments as he nursed it round the visual pattern at night and got it down again. I was taxiing out and heard it all over the VH radio—he sure sounded worried. Who wouldn't be?

Personally, I never enjoyed hand-flying the Shorts 360 and was always glad when one of the few fitted with an autopilot was available. The slab-sided box section fuselage made landing more difficult in a strong crosswind, and with its flat bottom fuselage it tended to float. One copilot applied full aileron the wrong way on a crosswind landing and the other wingtip nearly hit the ground before I could intervene. The old Dart Heralds were much better in crosswinds, you could put them down with a straight thirty knot crosswind as smoothly as 'a cat pissing on velvet'.

In overnight mail and parcels services it was not unusual to make four or even more, instrument approaches and landings in one duty period, and frequently in some quite foul winter weather. All good practice of course, but when some of our younger pilots tried to move up to the major airlines, nobody ever seemed interested in how many landings they'd done. Grand total of flying hours was always the thing. Yet, in some cases we were doing more landings in a week than some of the long-haul boys were doing in a year. Indeed, when the later jumbos and Airbuses came with enough range to make London to Singapore in one hop, flight duty regulations required three or even four pilots. The skipper would do the take off and the landings and the other two guys would get one each. Many years ago, there was the celebrated lawsuit involving a British Airways pilot who was fired for failing a check flight. He sued for unlawful dismissal showing that he'd only had just fifteen minutes in a pilot seat since his previous check. Thankfully he won the case.

Some captains liked to do rolling take-offs, thinking it would be easier on the passengers. Mindful of my experience with Britannia and Herald propeller systems, years before, I utterly refused to do that. One crew had trouble on the take-off run at Norwich which would not have happened if they had not done a rolling take-off. They got away with it, but if they'd tried the same thing at Plymouth's short runway, they'd have been off the end, down the bank, and probably killed. Same with intersection take-offs, that is, not using the full length of the runway. In many cases it is perfectly lawful to do so, but I never did. I always figured that if I had done so and even walked away from a take-off accident, then the 'wise after the event' types would all say yes, it was lawful, but it would have been better airmanship to take the full length. The extra taxiing distance was well worth it in my view. The three most useless things in aviation are runway behind you, altitude above you, and fuel burnt.

For some years we did carry passengers and one trip I pulled was a football charter. We had a full load of thirty-nine passengers so, when the agent came out with the load sheet, it was clear we were over maximum weight. Despite all the whingeing and threats from the group leader, there was no way I was going to break the law. Anyway, I couldn't see why there were so many in the group when a football team is only about a dozen players. I suppose the rest were the press and various hangers on. In the end, I offered them a choice. Either some of them stayed behind, or they all went back into the terminal and were individually weighed instead of taking the standard figure. I reckoned we could get away with that because the standard figure includes about 20Kg for baggage, and they had almost none. It worked, everyone was loaded, and we did the trip. They lost the game though.

In the end, we had to quit hauling passengers, which suited me fine because cargo is a lot less trouble. When the UK got into bed with the EU, we had to comply with all the silly rules the EU Authorities kept dreaming up. Our aeroplanes were fitted with TCAS collision avoidance equipment which we had been trained to use and been using for months. Then the EU stepped in, saying we had American sets which were not approved for use in Europe. The cost of fitting the Euro-kit was so expensive we could never recoup it within the expected life of the aircraft. Luckily, there was an opt out allowing us to operate without it, so long as we reduced seating capacity from 39 to 30. We did, but it lost us so much business that the management decided to quit quoting for passenger work altogether. Not that I minded—one of the buggers once stole my uniform cap. Thankfully, the rule only applied to passenger aeroplanes, which prompts one to wonder if they felt it was okay for cargo aeroplanes to collide?

In fact, the whole Euro-project has been a disaster for commercial aviation. About ten years after the UK joined the JAA, it had cost the civil aviation industry millions, and everyone had to admit it had done nothing whatsoever to improve air safety. When the idea was first mooted, most airline managements seem keen because of the opportunities it would open up in Europe. They seemed quite unaware that there were plenty of European based airlines itching to come compete for business here, and usually with lower overheads and costs.

It's been especially disastrous for UK trained professional pilots. Europeans learn enough aviation English to scrape by, and then get hired in the UK. Often, in their own countries, all the company technical manuals are written in their own language, so unless a Brit happens to speak that language there isn't much chance of getting hired. Of course, it's so often argued that Brits should pay more attention to learning foreign tongues but then the question becomes which one, or how many? Learning Italian wouldn't get anyone a job in Finland. Learning Portuguese would be no help applying for a job in Poland or vice versa. One friend of mine was offered a job flying a jet based in Italy. He wasn't able to accept because he had children of school age and the Italian schools could not take them, for they spoke no Italian.

A few years before I retired the CAA bought in the requirement for 'Cockpit Resource Management' training. I did quite well on the course, probably because I had enough experience to see through it all and give the answers they obviously wanted, but although much of it made a lot of sense, I was very cynical about the rest. After all, ultimately, the captain is responsible at law for the safe conduct of the flight, and you can't have an aeroplane flown by a committee. On the other hand, I think it is vital for a captain to listen to what a copilot has to say: a young first officer just out of training may know of something that the captain hasn't heard about yet or has maybe forgotten. Perhaps a more useful idea than CRM training would be to encourage airline pilots to read as many books as possible about past accidents.

Modern airline training puts a lot of emphasis on 'Standard Operating Procedures', which is fine, except for the simple fact that nobody can design a cut and dried checklist procedure to cover every possible emergency. Especially in cases where there is a compound emergency. I know of one accident years ago where a burst tire on take-off went un-noticed until the pilots retracted the gear, and the cabin began to fill up with smoke. The tire was blazing merrily away under the cabin and all the pilots had to do was put the gear down again. Climbing was no problem, all four engines were running fine, and the pilots followed the emergency checklist perfectly. But by the time they were back on final approach and did select the gear down again the fire had really taken hold. All on board died in the crash.

Sometimes it's a combination of background knowledge, experience, common sense and maybe sheer animal cunning that gets you out of a corner or, better still, prevents you getting into the corner to start with. One piece of advice I would give to any aspiring airline pilot is to read as many old accident reports as you can, as well as the many books that have been written about past accidents. Learn from the experience of others.

Autopilots don't get tired and don't get bored. No pilot can outfly a properly maintained autopilot and many modern airliners come with three of them fitted as standard. Modern jetliners are really designed to be operated via the autopilots, rather than hand flown. Manual flying, especially straight and level in the cruise, is like riding a bicycle. Once learnt, you cannot un-learn it and I always found long spells of hand flying tedious. I felt that the work of the grey cells was more profitably utilised in mulling over possible courses of emergency action, monitoring radios, instruments and weather, perhaps monitoring the other pilot and hoping they were monitoring me at the same time.

When I was just a few months away from retirement, Royal Mail had a major reorganisation programme and decided to use jets on all their overnight mail sectors. Consequently, we began losing contracts. Aircraft were disposed of and crews made redundant although, as the most senior non-management pilot, I was the last to go. Then they asked me to come back because the firm had been taken over by another. I returned after a couple of weeks leave, but over that summer the whole affair degenerated into chaos.

The final blow came when some clown at head office came up with the bright idea of demoting all over-sixty captains back to the right seat. There had never been any legal requirement to do that with aeroplanes of the type, and legally they would have had to continue paying us at the same rate. They were short of captains anyway, and hardly had any copilots ready for promotion. They also knew that within a year a law change was coming in to make that type of age discrimination unlawful. By then I had had enough so I told them to shove it.

Perhaps I was influenced by one of the last trips I did. We were heading north to RAF Kinloss and the weather up there was not looking good at all. Glasgow and Aberdeen were socked in, Inverness was closed overnight, and Edinburgh wasn't looking good either. The company were sending messages urging me to try for Kinloss, but I had other ideas. I didn't fancy two attempts at landing, and then having to divert to Prestwick. You could always rely on Prestwick being open, but I'd have barely had enough fuel to make it. Later, our chief pilot assured me I never would have made it to Kinloss. It was unusual weather for northern Scotland in August: a thunderstorm sat on top of the airfield for hours and the runway was fogged in at the same time. He'd been in a taxi heading for Kinloss to meet us but with flooding and mud slides he didn't think he was going to get there—even by road. I got into Edinburgh all right, but only just.

Chapter 27 - End of the Line

In the end, I was glad to be away from it all. Airlines will tell you that they always comply with the 'Highest International Regulations' but, as aircrew usually know, most of them adhere to the lowest standard they think they can consistently get away with! I'd flown just over 20,000 hours and worked for over twenty companies. I had been made redundant eight times, six firms I'd worked for had gone under leaving me out of a job, and I'd been fired twice. Quite a career!

When it was all over, I was awarded the Master Air Pilot certificate of the Guild of Air Pilots and Navigators, signed by HRH Prince Andrew. It was very kind of them, as I was not even a member of the Guild. It involved attending a black-tie dinner in London with many of aviation's great and good. A most enjoyable and interesting evening.

As I approached retirement, I thought it would be rather nice to revalidate my instructing qualifications and just do two or three days a week for as long as I could keep my medical status. Unfortunately, the basic flying training industry had taken such a hammering and lost so much business over the last few years, largely due to all the unnecessary new European rules and regulations. The cost/benefit analysis just didn't add up and I let all my UK licences lapse. For a few years I kept my American ATP current by flying a seaplane for a few hours in California each year but, in the end, I let that go too.

Master of the Guild of Air Pilots, Captain Peter Bugge, presents me with a Master Air Pilot Certificate, signed by prince Andrew, 2004.

The aviation world certainly changed a lot in the years I was involved, but most of the major changes took place in the first fifteen years. Apart from detail, not many major changes have happened since. Between 1957 and about 1970 Dragon Rapide biplanes had given way to executive jets, and most of the old propeller airliners had given way to jets, with Dakotas giving way to Boeing 737s. Crews were changing too. First to go were the Radio Officers, replaced by VHF and Single Sideband HF voice communication. Then it was the turn of the navigators, upstaged by Inertial Navigation systems and, eventually, GPS Satnav, with few of us left by 1980. Eventually, electronics replaced the Flight Engineers too. Concordes came and went, but there seems to be little chance of any more passenger supersonics. The fastest airliners now fly at about eighty-five percent of the speed of sound, and it seems likely it will stay that way. The most prolific of all jetliners, the Boeing 737, looks pretty much the same as it first did forty-five years ago. Boeing's claim that they still hope to be building 'Triple Sevens' in fifty years' time sounds credible.

Helicopters are something I know very little about and I only ever flew in one once, a ten-minute hop around

Manhattan. From the moment the pilot started the engine the whole thing seemed to be trying to shake itself apart, so I never fancied another shot, although many of my colleagues learned to fly them, or wanted to. I always had the greatest admiration for the boys flying them out in the North Sea, some of the toughest commercial flying anywhere. Helicopters are all very complex machines and all the former rotary wing pilots I flew with seemed to adapt to fixed wing flying very well.

At one stage, I had a crack at gliding, which is an incredibly enjoyable form of flight, but I found gliding clubs to be rather poorly organised. Compared with power flight instructors, I found gliding instructors very keen but amateurish. Preflight briefings were usually done in a matter of seconds, then the instructor would spend ten minutes after a five-minute flight telling you what you'd done wrong. In the end, I gave it up without even going solo. I did do a few air tows with a Piper Cub and that was quite a lot of fun.

I had qualified as a commercial seaplane pilot but never managed to get a job in that side of the business. That's always seemed to me one of the most attractive ways of making a living, but it's mostly incredibly hard work. Once you are in the air it's the easy bit— just flying an aeroplane a bit slower than usual, because of the extra drag from the floats. Even landing and taking off I never found too demanding, although you have to remember a seaplane has no shock absorbers. The difficult part is manoeuvring on the water and docking, especially since you have no brakes.

Flying has become the safest form of travel thanks to technological developments leading to reliable communication, better airports and airport lighting, more accurate weather forecasts, better navigation systems and mayb,e most of all, more reliable engines, although the downside of that is the extensive use of twin-engined airliners on very long overwater sectors. Indeed, every year more people actually curl up and die comfortably strapped into their passenger seats than are killed in airline accidents worldwide.

As for those long overwater sectors in twin-engined aircraft, I'm never very happy as a passenger, but there isn't much choice now. Most North Atlantic flights are in twins. One twin has already made an emergency glide approach into the Azores and that never could have happened to a four-engined aircraft like a Boeing 707, with a flight engineer and seven, rather than just three, fuel tanks. After all, there are more aeroplanes at the bottom of the sea than there are submarines in the sky!

'Open Skies' deregulation and the 'low fare revolution' may have reduced the cost of air travel but, at the same time, caused massive instability in the airline industry. The public love the cheap tickets, but they're the first to complain when they get stranded halfway across the world because the airline has gone down the tubes. Gone are the days when pilots could look forward to a reasonably secure career, and that applies equally to engineers, office, catering staff and everyone else. 'New start' airlines appear all the time, but so few of them last for very long. Right now, I can think of only two UK airline companies that are still trading under their original names and have been for much more than twenty-five years. It has become increasingly difficult for youngsters to get into the business, despite their massive financial investments in their training. As seat demand grows, airliners get bigger and, to some extent, that means fewer pilot vacancies rather than more.

We have heard a lot in recent years about a third runway at Heathrow, notwithstanding the fact that it's not so many years since they closed cross runway 05/23 because it was hardly ever used. If ever they do make the mistake of ripping a large slice of Middlesex up to build it, I guess it will be in service just in time for the world to run out of jet fuel. Then there is all the talk about a second runway at Gatwick, despite the fact that there are actually two there already, although I think the other one is limited to twenty days a year use, and visual approaches only. When I was working for the CAA, the Deputy Director of Aerodrome Standards himself demonstrated to me that it was totally impossible to site a major runway anywhere near the existing ones and still comply with international runway design regulations. Even if it were possible, they'd have to demolish Charlwood Hill, which has some of the most expensive real estate in the land. There is still enough real estate available to build a second runway at Stansted but tearing apart so much beautiful English countryside so that a few foreign airlines can make even more money would be a monstrous sacrilege.

As to the future, there are the optimists who believe some new form of propulsion will eventually replace the jet engine but of that I have my doubts. Technological progress does not develop in an exponential, or even linear manner. It has been argued that from the fall of the Roman Empire to the time of American Independence there were few new inventions other than the printing press, gunpowder, and the optics that gave us the telescope, microscope and reading glasses. The next century gave us a mass of new inventions but in the twentieth, the only completely new inventions were the jet engine, television, the microchip, the unravelling of the DNA code, and the Nuclear Bomb. Flying belongs to the nineteenth century, a few gliders having flown before the Wright Brothers powered flight success. Can we really assume that major inventions will continue to flow, indefinitely, from the minds of the brilliant?

Someday there may be sub-orbital rocket planes that can make Sydney in four hours. Maybe someday there'll be chocolate bars and potato chips that make you lose weight. But don't hold your breath waiting!

Do I miss flying? No, not really. If I had kept current, I would only be able to rent the more mundane and uninteresting flying club machines. Nobody would be likely to let me fly any interesting antique aeroplanes, and in any case, I wouldn't want to because it wouldn't happen often enough to remain competent, and I surely don't want to be the guy who wrecks some lovely old machine the engineers have just spent ten years restoring. Antique warbird aircraft have a quite bad safety record, despite the high quality maintenance they certainly have, and despite the high experience levels of those fortunate to fly them. I suspect the problem is, to some extent, that they don't really get flown enough and the pilots don't get to fly them enough. A fellow former Dart Herald captain, the late John Fairey, was a highly esteemed air show pilot with extensive warbird experience, yet he was killed when the engine just fell of a Percival Provost he was flying straight and level. If that can happen to one of the stars of the business, then we ordinary mortals aren't even in with a chance. Anyway, I got into the business to travel and make money, not get rid of it.

I keep busy. My photo collection has developed into a kind of picture agency and I spend a lot of time supplying photos to authors, researchers, and aircraft owners. I drive hospital cars and I spend a couple of months a year in Australia, New Zealand and the USA. We have two very demanding cats to take care of as well!

Over the years I had visited every single state of America, including driving the Alaska Highway both ways. I have explored much of New Zealand and Canada and, altogether, visited seventy-eight countries—most of which I would not care to return to! Nevertheless, I had spent little time in Australia. So, with no restrictions upon my time away from home, I have been able to drive the Stuart Highway and spend time in every Australian state, and there is always something around the house that needs doing.

Would I do it all again? Well, I guess I would like to if the whole environment were the same as in the sixties and seventies, but certainly not the way it is nowadays. In some ways I guess I was extremely lucky to have been born in England at the time I was, and in one of the few countries in the world where a kid from the 'wrong side of the tracks', who wore glasses and left school with no money and four useless 'O levels', could eventually make it to airline captain. Many things have changed, mostly for the worse, and it would be a lot more difficult for a kid to do that these days. Flying is great … it's getting up there that's the problem!

Appendix

Licences

British Airline Transport Pilot Licence (rated for Handley Page Herald, Short 360, Beech King Air, Cessna Golden Eagle, Piper Navajo).
Instructor and Examiner rating for all singles and multis up to 12,500 lbs max take off weight.
American Air Transport Licence (including single engine seaplanes).
British and American Flight Navigator Licences.
American FAA Flight Dispatcher Licence.
American FAA Airframe and Powerplant Licences.
American FAA Advanced and Instrument Ground Instructor Licences.
Associate degree in Navigation Science & technology.

Types flown as navigator

Bristol Britannia, Douglas DC-8, Canadair 44, Boeing 707, Vickers Viscount, Hawker Siddeley 748.

Types flown as pilot

Nearly a hundred!

Total hours

Pilot: about 11,400.
Navigator: about 8,600.

Glossary

ADF, Automatic Direction Finding. A marine or aircraft radio-navigation instrument that automatically and continuously displays the relative bearing from the ship or aircraft to a suitable radio station.

ATC, Air Traffic Control

Consol, Originally known as Sonne. A radio navigation system developed in Germany during World War Two which remained in limited use until 1991.

CPL, Commercial Pilot's Licence

DEW Radar, Distant Early Warning Line. Also known as the DEW Line or Early Warning Line. A system of radar stations in the northern Arctic region of Canada, with additional stations in Alaska, Faroe Islands, Greenland, and Iceland.

DME, Distance Measuring Equipment. A navigation beacon, usually coupled with a VOR beacon, which enables aircraft to measure their position relative to that beacon. Aircraft send out a signal which is sent back after a fixed delay by the DME ground equipment.

Doppler Radar, A specialized radar that uses the Doppler effect to produce velocity data about objects at a distance. It does this by bouncing a microwave signal off a desired target and analysing how the object's motion has altered the frequency of the returned signal.

GNSS, Global Navigation Satellite System. Any satellite constellation that provides positioning, navigation, and timing services on a global or regional basis.

GPS, Global Positioning System. Sometimes confused with GNSS, GPS is a satellite-based radio navigation system owned by the United States government and operated by the United States Space Force. Most modern satellite navigation systems in ships and aircraft can interrogate a range of satellite constellations, including GPS.

Gyrocompass, A type of non-magnetic compass which is based on a fast-spinning disc and the rotation of the Earth to find geographical direction.

ILS, Instrument Landing System. A precision runway approach aid based on two radio beams which together provide pilots with both vertical and horizontal guidance during an approach to land.

INS, Inertial Navigation System, a self-contained navigation system in which measurements provided by accelerometers and gyroscopes are used to track the position and orientation of an object relative to a known starting point.

IR, Instrument Rating. An endorsement which permits a pilot to fly the aircraft solely with reference to the in-cockpit instrumentation.

LORAN, Long Range Navigation. A hyperbolic radio navigation system developed in the United States during World War Two. Widely used by ships and aircraft until the 1970s, it remained in limited use until 1991.

NOTAM, Notice to Airmen. A notice published by an aviation authority to alert pilots of potential hazards along a flight route or at a location that could affect flight.

OTS, Organised Tracking System. The North Atlantic Organised Track System (NAT-OTS), is a structured set of transatlantic flight routes that stretch from eastern North America to western Europe across the Atlantic Ocean.

Transponder, Transponders give information to ATC about an aircraft's location in space and, in most cases, its altitude as well.

VHF, Very High Frequency. VHF radio is widely used in aviation for both voice communications and navigation.

VOR, VHF omnidirectional range. A short-range navigation aid which gives a pilot information on direction and range from a ground station.

NOTE, In the United States, the FAA has decided that, in future, GPS will be used as virtually the sole source for all aircraft navigation. Many ground-based navigational aids will be de-commissioned by 2030. This includes ILS, VOR, DME, and NDB.

SunRise

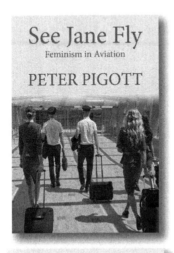

See Jane Fly

Feminism in Aviation

PETER PIGOTT

Sky Talk

Stories from
flying's Golden
Age

Philip Hogge

The Golden Age
of Flying Boats

Peter Pigott

THE CONSTELLATION
Lockheed's Graceful Masterpiece
Alexander Clifton

TWA

WORLD AIR LINES

www.sunpub.info

Printed in Great Britain
by Amazon